SpringerBriefs in Mathematics

SpringerBriefs present concise summaries of cutting-edge research and practical applications across a wide spectrum of fields. Featuring compact volumes of 50 to 125 pages, the series covers a range of content from professional to academic. Briefs are characterized by fast, global electronic dissemination, standard publishing contracts, standardized manuscript preparation and formatting guidelines, and expedited production schedules.

Typical topics might include:

A timely report of state-of-the art techniques A bridge between new research results, as published in journal articles, and a contextual literature review A snapshot of a hot or emerging topic An in-depth case study A presentation of core concepts that students must understand in order to make independent contributions

SpringerBriefs in Mathematics showcases expositions in all areas of mathematics and applied mathematics. Manuscripts presenting new results or a single new result in a classical field, new field, or an emerging topic, applications, or bridges between new results and already published works, are encouraged. The series is intended for mathematicians and applied mathematicians. All works are peer-reviewed to meet the highest standards of scientific literature.

Titles from this series are indexed by Scopus, Web of Science, Mathematical Reviews, and zbMATH.

More information about this series at https://link.springer.com/bookseries/10030

Rui A. C. Ferreira

Discrete Fractional Calculus and Fractional Difference Equations

Springer

Rui A. C. Ferreira
Group of Mathematical Physics
University of Lisbon
Lisbon, Portugal

ISSN 2191-8198 ISSN 2191-8201 (electronic)
SpringerBriefs in Mathematics
ISBN 978-3-030-92723-3 ISBN 978-3-030-92724-0 (eBook)
https://doi.org/10.1007/978-3-030-92724-0

This Springer imprint is published by the registered company Springer Nature Switzerland AG
The registered company address is: Gewerbestrasse 11, 6330 Cham, Switzerland

To my family

Preface

It is historically accepted that the fractional calculus was born when, in 1695, L'Hôpital inquired Leibniz about what meaning could be ascribed to $d^n f(x)/dx^n$ if n were a fraction. Since that time, the fractional calculus has drawn the attention of famous mathematicians, such as Euler, Laplace, Abel, Liouville, and Riemann. But it was not until 1884 that the theory of fractional operators achieved a level in its development suitable as a point of departure for the modern mathematician [47]. Nowadays the theory includes operators D^w, where w can be any complex number. Therefore, the nomenclature *fractional calculus* became somewhat a misnomer, though it is commonly accepted by the scientific community.

The discrete calculus together with the study of difference equations[1] is in these times a well-established theory (see, e.g., [19, 41]). It is used, for example, in the modelling of economical and biological phenomena.

The primary goal of this book is to merge these two theories, i.e., the fractional calculus and the discrete calculus, in a concise but comprehensive way. It is designed primarily for graduate students notwithstanding it may be useful for any researcher interested in the theory of discrete fractional calculus and fractional difference equations. This book is written in a self-contained manner and requires no particular background from a student, though some knowledge on calculus and/or differential/difference equations is an asset.

In 2015 it was published the first book within the topic "discrete fractional calculus and fractional difference equations" [35]. An extensive collection of results is presented therein with particular emphasis on the *nabla* or *backward* fractional operators. In this book we present a thorough study of the *delta* or *forward* fractional calculus and related fractional difference equations. We should mention that our approach is substantially different from the one in [35], particularly when developing the theory of fractional difference equations. In [35] it is heavily used the (discrete) Laplace transform technique in order to solve fractional difference equations, whereas, here, we use only techniques emerging from the initially

[1] In this book we will use the words "discrete" or "difference" indiscriminately.

developed (in previous chapters) discrete fractional calculus theory. In this sense, one may say that this book and the book by Goodrich and Peterson [35] complement each other and an interested reader will find plenty (and different) information in both.

The process of writing this book was inspired by the works listed in the References section [1–50]. This book comprises four chapters, which are then divided into several sections and possibly subsections. We describe its content succinctly in the following.

In Chap. 1 we develop the basic forward (or "delta" Δ) discrete calculus. We define the difference and summation operators and present their most useful properties. We should mention in passage that throughout the text we have tried to be as thorough as possible when presenting the proofs of our results. Finally, we define the discrete exponential function as it will be used many times within this work.

In Chap. 2 we develop the mathematical theory of the discrete fractional calculus. We introduce the "left" and "right" fractional operators, and subsequently, we prove many essential results such as power rules, composition rules, etc. It should be emphasized that the power rules appearing in Sect. 2.3 emerge from a different proof than the one presented in [35, Section 2.4]. We follow more closely the proof presented in [36]. Still in this chapter, we derive the Saalschutz formula as an application of the fractional Leibniz formula, and we show how fractional operators may be used to define novel entropic functionals (that may, in turn, be useful for physicists).

In Chap. 3 we collect to a great extent the author's work in discrete fractional linear equations within the past 8 years [21, 23]. In particular we provide here the explicit solution of the linear fractional initial value problem of order $0 < \alpha \leq 1$. We also present some results on the asymptotic behavior of solutions and the stability results of fractional difference systems obtained in [16]. With that respect, we have marked the section entitled "Stability" with an $(*)$ because we omitted the proof of the main result therein presented. The reason is that it makes use of the z-transform (which is equivalent to the discrete Laplace transform under a certain transformation) and complex analysis techniques which, as already mentioned above, we wanted to exclude from our analysis. We nevertheless opted to include Sect. 3.1.3 in this work, as we believe that the stability results therein presented are of fundamental and independent interest for researchers studying fractional difference equations. Finally, we study the influence of perturbed data in nonlinear equations and some boundary value problems.

In Chap. 4 we present and develop the foundational results of the discrete fractional calculus theory. We consider the basic problem of the calculus of variations with fixed and non-fixed boundary conditions, and we derive first- and second-order optimality conditions, as well as a sufficient condition of optimality.

We included at the end of each chapter a section containing exercises. We believe that this will benefit the readers of this book, in particular the students who will attend a course on discrete fractional calculus.

Finally, we should like to mention that we hope this book has sufficient virtues so that a reader would like to keep it around. We expect that the techniques of the discrete fractional calculus which we present will add useful tools to the researcher's repertoire of methods for attacking discrete analytical problems.

Acknowledgments Rui A. C. Ferreira was supported by the "Fundação para a Ciência e a Tecnologia (FCT)" through the program "Stimulus of Scientific Employment, Individual Support-2017 Call" with reference CEECIND/00640/2017.

Lisbon, Portugal
March 2021

Rui A. C. Ferreira

Contents

1 Discrete Calculus 1
1.1 The Difference and Summation Operators 1
1.2 Discrete Exponential Function 9
1.3 Exercises 13

2 Discrete Fractional Calculus 15
2.1 Motivation 15
2.2 The Fractional Sums and Differences 17
2.3 Power Rules 22
2.4 Composition Rules 27
2.5 Fractional Leibniz Formula 30
2.6 A Digression into Physics: A Novel Entropic Functional 35
2.7 Exercises 39

3 Fractional Difference Equations 41
3.1 Linear Difference Equations 42
3.1.1 Two Inequalities 48
3.1.2 Asymptotic Behavior 53
3.1.3 Stability* 60
3.2 Nonlinear Difference Equations: Influence of Perturbed Data 63
3.3 Boundary Value Problems 66
3.4 Exercises 69

4 Calculus of Variations 71
4.1 Necessary Conditions 73
4.2 Natural Boundary Conditions 77
4.3 A Sufficient Condition 80
4.4 Exercises 81

References 83

Index 87

Chapter 1
Discrete Calculus

This chapter is a fairly compelling introduction to the basic notions and notations of *discrete forward*[1] *calculus*.

1.1 The Difference and Summation Operators

We start by introducing some notation that in turn will be used throughout this text. For a number $a \in \mathbb{R}$, we put

$$\mathbb{N}_a = \{a, a+1, \ldots\} \text{ and } \mathbb{N}^a = \{\ldots, a-1, a\}.$$

Sometimes we will also write $\mathbb{N}_a^b = \{a, a+1, \ldots, b\}$, where $b = a+k$ with $k \in \mathbb{N}_1$.

Definition 1.1 Consider a function $f : \mathbb{N}_a \to \mathbb{R}$. The forward difference operator is defined by $\Delta[f](t) = f(t+1) - f(t)$, for $t \in \mathbb{N}_a$. Also, we define higher-order differences recursively as $\Delta^n[f](t) = \Delta[\Delta^{n-1} f](t)$, $n \in \mathbb{N}_1$, where Δ^0 is the identity operator, i.e., $\Delta^0[f](t) = f(t)$.

Remark 1.2 We will often drop the use of the bracket $[\cdot]$ when using operators as long as one understands the true meaning of the notation.

We list several important properties of the forward difference operator in the next theorem.

[1] The word "forward" will become clear in Definition 1.1. We will drop the use of this word eventually within this book.

R. A. C. Ferreira, *Discrete Fractional Calculus and Fractional Difference Equations*, SpringerBriefs in Mathematics,
https://doi.org/10.1007/978-3-030-92724-0_1

Theorem 1.3 *Assume that $f, g : \mathbb{N}_a \to \mathbb{R}$ and $\beta, \gamma \in \mathbb{R}$. Then, for $t \in \mathbb{N}_a$, we have*

1. $\Delta[\beta](t) = 0$.
2. $\Delta[\beta f](t) = \beta \Delta f(t)$.
3. $\Delta[f + g](t) = \Delta f(t) + \Delta g(t)$.
4. $\Delta[\beta^{s+\gamma}](t) = (\beta - 1)\beta^{t+\gamma}$.
5. $\Delta[fg](t) = \Delta f(t)g(t) + f(t+1)\Delta g(t)$.
6. $\Delta\left[\frac{f}{g}\right](t) = \frac{\Delta f(t)g(t) - f(t)\Delta g(t)}{g(t)g(t+1)}$, *if* $g(t)g(t+1) \neq 0$.

Proof The proofs are straightforward, so we prove 4. and 5. as examples.

We have

$$\Delta[\beta^{s+\gamma}](t) = \beta^{t+1+\gamma} - \beta^{t+\gamma} = (\beta - 1)\beta^{t+\gamma},$$

and

$$\begin{aligned}\Delta[fg](t) &= f(t+1)g(t+1) - f(t)g(t) \\ &= f(t+1)[g(t+1) - g(t)] + f(t+1)g(t) - f(t)g(t) \\ &= f(t+1)\Delta g(t) + \Delta f(t)g(t),\end{aligned}$$

which prove the claims.

Example 1.4 Consider the function $f(t) = t^2$ defined on $\mathbb{N}_a$, for a real number a. Then, $\Delta f(t) = t + t + 1 = 2t + 1$. Notice that, contrarily to the *continuous calculus case*, in this context $\Delta f(t) \neq 2t$.

Consider the Gamma function[2] $\Gamma(z) = \int_0^\infty e^{-t}t^{z-1}dt$, Re $z > 0$. It satisfies the following formula:

$$\Gamma(z+1) = z\Gamma(z), \tag{1.1}$$

which will be used repeatedly in this work. We remark that equality (1.1) permits us to say that the Gamma function generalizes the factorial function, as $\Gamma(n+1) = n!$ for $n \in \mathbb{N}_0$. Moreover, we use (1.1) to extend the domain of Γ. Indeed, as in this work we will be dealing only with real numbers, we define

$$\Gamma(x) = \frac{\Gamma(x+1)}{x}, \quad x \in \mathbb{R}^- \backslash \mathbb{N}^0.$$

---

[2] A good account of formulas related with the gamma function may be consulted, e.g., in [47, Appendix B].

Definition 1.5 (Falling Function) The falling function is defined, for $x, y \in A \subset \mathbb{R}$, by

$$x^{\underline{y}} = \begin{cases} x(x-1)\dots(x-y+1) \text{ for } y \in \mathbb{N}_1, \\ 1 \text{ for } y = 0, \\ \frac{\Gamma(x+1)}{\Gamma(x+1-y)} \text{ for } x, x-y \notin \mathbb{N}^{-1}, \\ 0 \text{ for } x \notin \mathbb{N}^{-1} \text{ and } x-y \in \mathbb{N}^{-1}. \end{cases}$$

The falling function *acts* within the discrete calculus theory as the power function does within the *continuous* calculus theory. Indeed, this may be seen in the content of the following result.

Proposition 1.6 (Power Rules) *Let α and r be some constants. Then, the following equalities:*

$$\Delta[(s+\alpha)^{\underline{r}}](t) = r(t+\alpha)^{\underline{r-1}}, \tag{1.2}$$

and

$$\Delta[(\alpha-s)^{\underline{r}}](t) = -r(\alpha-(t+1))^{\underline{r-1}}, \tag{1.3}$$

hold, whenever the expressions in these two formulas are well-defined (cf. Definition 1.5).

Proof We will prove (1.2) leaving the other one to the reader.

If $r = 0$, then the result is obvious. So, assume $r \neq 0$. We have

$$\Delta[(s+\alpha)^{\underline{r}}](t) = (t+1+\alpha)^{\underline{r}} - (t+\alpha)^{\underline{r}}.$$

If $r \in \mathbb{N}_1$, then

$$\begin{aligned}\Delta[(s+\alpha)^{\underline{r}}](t) &= (t+1+\alpha)(t+\alpha)\dots(t+2+\alpha-r) \\ &\quad - (t+\alpha)(t+\alpha-1)\dots(t+2+\alpha-r)(t+\alpha-r+1) \\ &= r(t+\alpha)^{\underline{r-1}}.\end{aligned}$$

If $r \notin \mathbb{N}_1$, then it is easy to check the case where $t+\alpha-r+1 \in \mathbb{N}^{-1}$ and $t+\alpha \notin \mathbb{N}^{-1}$. It remains to prove the case where $t+\alpha, t+\alpha-r+1 \notin \mathbb{N}^{-1}$; we have

$$\begin{aligned}\Delta(t+\alpha)^{\underline{r}} &= \frac{\Gamma(t+2+\alpha)}{\Gamma(t+2+\alpha-r)} - \frac{\Gamma(t+1+\alpha)}{\Gamma(t+1+\alpha-r)} \\ &= \frac{(t+1+\alpha)\Gamma(t+1+\alpha) - (t+1+\alpha-r)\Gamma(t+1+\alpha)}{\Gamma(t+2+\alpha-r)} \\ &= r\frac{\Gamma(t+\alpha+1)}{\Gamma(t+\alpha+1-(r-1))} = r(t+\alpha)^{\underline{r-1}},\end{aligned}$$

and the proof is done.

Remark 1.7 It is easily seen that the following formula holds:

$$t^{\underline{\alpha+\beta}} = (t-\beta)^{\underline{\alpha}} t^{\underline{\beta}}. \tag{1.4}$$

We now observe that when $n \geq k \geq 0$ are integers, then the binomial coefficient $\binom{n}{k} = \frac{n!}{k!(n-k)!}$ satisfies

$$\binom{n}{k} = \frac{n(n-1)\dots(n-k+1)}{k!} = \frac{n^{\underline{k}}}{\Gamma(k+1)}. \tag{1.5}$$

Motivated by (1.5), we also put

$$\binom{t}{r} = \frac{t^{\underline{r}}}{\Gamma(r+1)}, \tag{1.6}$$

even for noninteger values of t and r, where $\binom{t}{r} = 0$ if $r \in \mathbb{N}^{-1}$ and $t^{\underline{r}}$ is defined.

A related concept with the falling function is the *rising function* or the *Pochhammer symbol.*

Definition 1.8 The Pochhammer symbol is defined, for $x, y \in A \subset \mathbb{R}$, by

$$(x)_y = \begin{cases} x(x+1)\dots(x+y-1) \text{ for } y \in \mathbb{N}_1, \\ 1 \text{ for } y = 0, \\ \frac{\Gamma(x+y)}{\Gamma(x)} \text{ for } x, x+y \notin \mathbb{N}^0, \\ 0 \text{ for } x \in \mathbb{N}^0 \text{ and } x+y \notin \mathbb{N}^0. \end{cases}$$

It is clear that

$$(t+\alpha-1)^{\underline{\alpha}} = (t)_\alpha,$$

and

$$(-1)^n = \frac{(-\alpha)^{\underline{n}}}{(\alpha)_n}, \quad n \in \mathbb{N}_0. \tag{1.7}$$

Before we proceed to define the summation operator, we present the (generalized) Leibniz formula and a chain rule, both arising from the discrete calculus theory.

Theorem 1.9 (Leibniz Formula) *Consider two functions $f, g : \mathbb{N}_a \to \mathbb{R}$ and $n \in \mathbb{N}_1$. Then,*

$$\Delta^n[fg](t) = \sum_{k=0}^{n} \binom{n}{k} \Delta^{n-k} f(t+k) \Delta^k g(t), \quad t \in \mathbb{N}_a. \tag{1.8}$$

Proof Fix $t \in \mathbb{N}_a$. We will use induction on $n \in \mathbb{N}_1$.

When $n = 1$, then (1.8) holds for any functions f and g by 5. of Theorem 1.3. Now, let $n \in \mathbb{N}_1$ and assume that (1.8) holds for functions f and g. Then,

$$
\begin{aligned}
\Delta^{n+1}[fg](t) &= \Delta[\Delta^n[fg]](t) \\
&= \sum_{k=0}^{n} \binom{n}{k} \Delta^{n+1-k} f(t+k)\Delta^k g(t) + \sum_{k=0}^{n} \binom{n}{k} \Delta^{n-k} f(t+1+k)\Delta^{k+1} g(t) \\
&= \sum_{k=0}^{n} \binom{n}{k} \Delta^{n+1-k} f(t+k)\Delta^k g(t) + \sum_{k=1}^{n+1} \binom{n}{k-1} \Delta^{n+1-k} f(t+k)\Delta^k g(t) \\
&= \Delta^{n+1} f(t) \cdot g(t) + \sum_{k=1}^{n} \binom{n+1}{k} \Delta^{n+1-k} f(t+k)\Delta^k g(t) \\
&\quad + f(t+n+1)\Delta^{n+1} g(t) \\
&= \sum_{k=0}^{n+1} \binom{n+1}{k} \Delta^{n+1-k} f(t+k)\Delta^k g(t),
\end{aligned}
$$

where we have used the well-known formula of the binomial coefficient:

$$\binom{n+1}{k} = \binom{n}{k} + \binom{n}{k-1}, \quad k \in \{1, \dots, n\}.$$

The proof is done.

Theorem 1.10 (Chain Rule) *Let $f : \mathbb{R} \to \mathbb{R}$ be a continuously differentiable function and consider $g : \mathbb{N}_a \to \mathbb{R}$. Then, we have*

$$\Delta[f \circ g](t) = \int_0^1 f'(g(t) + h\Delta g(t))dh\Delta g(t), \quad t \in \mathbb{N}_a.$$

Proof The reader just has to note that, after a substitution, we get

$$\int_0^1 f'(g(t) + h\Delta g(t))dh\Delta g(t) = \int_{g(t)}^{g(t+1)} f'(s)ds = f(g(t+1)) - f(g(t)),$$

and the proof is done.

Example 1.11 In order to illustrate the previous theorem, consider the functions

$$f(x) = e^x, \ x \in \mathbb{R}, \ \text{and} \ g(t) = t^2, \ t \in \mathbb{N}_0.$$

Since $\Delta g(t) = 2t + 1$, using Theorem 1.10, we get

$$\begin{aligned}\Delta[f \circ g](t) &= \int_0^1 e^{t^2+h(2t+1)} dh(2t+1) \\ &= (2t+1)e^{t^2} \int_0^1 e^{h(2t+1)} dh \\ &= e^{t^2}(e^{2t+1} - 1).\end{aligned}$$

Let us also calculate $\Delta[f \circ g](t)$ using the definition. We have

$$\begin{aligned}\Delta[f \circ g](t) &= f(g(t+1)) - f(g(t)) \\ &= e^{(t+1)^2} - e^{t^2} \\ &= e^{t^2+2t+1} - e^{t^2} \\ &= e^{t^2}(e^{2t+1} - 1).\end{aligned}$$

Definition 1.12 Let $a \in \mathbb{R}$ and $t \in \mathbb{N}_{a+1}$. For a function $f : \mathbb{N}_a \to \mathbb{R}$, we define the summation of f from a to $t - 1$ by

$$\Delta_a^{-1} f(t) = \sum_{s=a}^{t-1} f(s) = f(a) + f(a+1) + \ldots + f(t-1).$$

We assume that empty sums are zero, i.e., $\Delta_a^{-1} f(a) = \sum_{s=a}^{a-1} f(s) = 0$ for all $a \in \mathbb{R}$.

The following theorem provides some properties for the summation operator defined above.

Theorem 1.13 *Assume* $f, g : \mathbb{N}_a \to \mathbb{R}$, $b, c, d \in \mathbb{N}_a$ *with* $b \preceq c \preceq d$, *and* $\beta \in \mathbb{R}$. *Then,*

1. $\Delta_b^{-1}[\beta f](c) = \beta \Delta_b^{-1} f(c)$.
2. $\Delta_b^{-1}[f + g](c) = \Delta_b^{-1} f(c) + \Delta_b^{-1} g(c)$.
3. $\Delta_b^{-1} f(d) = \Delta_b^{-1} f(c) + \Delta_c^{-1} f(d)$.
4. $\left|\Delta_b^{-1} f(c)\right| \leq \Delta_b^{-1} |f|(c)$.

Proof We prove item 4 and leave the others to the reader.

Using the triangle inequality, we have

$$\left|\Delta_b^{-1} f(c)\right| = \left|\sum_{s=b}^{c-1} f(s)\right| \leq \sum_{s=b}^{c-1} |f(s)| = \Delta_b^{-1} |f|(c),$$

and the proof is done.

Theorem 1.14 (Fundamental Theorem of the Discrete Calculus) *Assume* $f : \mathbb{N}_a^{b-1} \to \mathbb{R}$ *and consider the function* $F : \mathbb{N}_a^b \to \mathbb{R}$ *defined by* $F(t) = \Delta_a^{-1} f(t)$. *Then,* $\Delta F(t) = f(t)$ *on* $\mathbb{N}_a^{b-1}$.

Proof For $t \in \mathbb{N}_a^{b-1}$, we have

$$\Delta F(t) = \sum_{s=a}^{t} f(s) - \sum_{s=a}^{t-1} f(s) = f(t),$$

and the proof is done.

Corollary 1.15 *If* $f : \mathbb{N}_a^{b-1} \to \mathbb{R}$ *and* F *is such that* $\Delta F(t) = f(t)$ *on* $\mathbb{N}_a^{b-1}$, *then* $\Delta_a^{-1} f(b) = F(b) - F(a)$.

Proof Let $G(t) = \Delta_a^{-1} f(t)$ on $\mathbb{N}_a^b$. By Theorem 1.14, $\Delta G(t) = f(t)$ on $\mathbb{N}_a^{b-1}$. Since $\Delta F(t) = f(t)$ on $\mathbb{N}_a^{b-1}$, we must have $\Delta[G - F](t) = 0$ for all $t \in \mathbb{N}_a^{b-1}$. Therefore, $G(t) - F(t) = c$ for some $c \in \mathbb{R}$ and all $t \in \mathbb{N}_a^b$. Hence,

$$F(b) - F(a) = (G(b) - c) - (G(a) - c) = G(b) = \Delta_a^{-1} f(b),$$

which concludes the proof.

Example 1.16 Let $f(t) = (t - \alpha)^{\underline{r}}$, with $\alpha \in \mathbb{R}$ and $r \notin \mathbb{N}^{-1}$. Put $F(t) = \frac{(t-\alpha)^{\underline{r+1}}}{r+1}$. Then, $\Delta F(t) = f(t)$. Hence, by Corollary 1.15 and the fact that $\frac{r^{\underline{r+1}}}{r+1} = 0$, we have

$$\Delta_{\alpha+r}^{-1} f(t) = \frac{(t-\alpha)^{\underline{r+1}}}{r+1}, \quad t \in \mathbb{N}_{\alpha+r}. \tag{1.9}$$

Note in particular that if $\alpha = 0$ and $r = 1$, we have just obtained the following well-known formula (for the triangular numbers):

$$\sum_{k=1}^{n} k = \frac{n(n+1)}{2}. \tag{1.10}$$

Example 1.17 Let $f(t) = r^t$ with $r \neq 1$. If $F(t) = \frac{r^t}{r-1}$, then $\Delta F(t) = f(t)$. By Corollary 1.15, we get

$$\sum_{s=0}^{t-1} r^s = \frac{r^t - 1}{r - 1}, \quad t \in \mathbb{N}_0,$$

which is the well-known formula used to obtain the geometric series.

Proposition 1.18 (Summation by Parts) *Given two functions $f, g : \mathbb{N}_a \to \mathbb{R}$ and $b, c \in \mathbb{N}_a$ with $b < c$, we have the summation by parts formula:*

$$\sum_{s=b}^{c-1} f(s)\Delta g(s) = [f(s)g(s)]_{s=b}^{s=c} - \sum_{s=b}^{c-1} \Delta f(s)g(s+1). \tag{1.11}$$

Proof The formula is a consequence of item 5. in Theorem 1.3 and Corollary 1.15.

Example 1.19 Let us find $\sum_{k=1}^{n} k^2$.

By using (1.9), we get

$$\sum_{k=2}^{n} k^{\underline{2}} = \frac{(n+1)^{\underline{3}}}{3} = \frac{(n+1)n(n-1)}{3}, \quad n \in \mathbb{N}_1.$$

Since $1^{\underline{2}} = 0$, then we have by Proposition 1.18 and with the help of (1.10)

$$\begin{aligned}\sum_{k=2}^{n} k^{\underline{2}} = \sum_{k=1}^{n} k^{\underline{2}} &= \left[kk^{\underline{2}}\right]_{k=1}^{k=n+1} - 2\sum_{k=1}^{n}(k+1)k \\ &= (n+1)(n+1)n - 2\sum_{k=1}^{n} k^2 - n(n+1) \\ &= (n+1)n^2 - 2\sum_{k=1}^{n} k^2.\end{aligned}$$

Therefore,

$$2\sum_{k=1}^{n} k^2 = (n+1)n^2 - \frac{(n+1)n(n-1)}{3},$$

or

$$\sum_{k=1}^{n} k^2 = \frac{(n+1)n(2n+1)}{6}.$$

We end this section by presenting a Leibniz type rule using the discrete operators studied so far.

Proposition 1.20 (Leibniz Rule) *Assume $f : \mathbb{N}_{a+1} \times \mathbb{N}_a \to \mathbb{R}$ and $g : \mathbb{N}^b \times \mathbb{N}^b \to \mathbb{R}$. Then,*

$$\Delta_t \sum_{s=a}^{t-1} f(t,s) = \sum_{s=a}^{t-1} \Delta_t f(t,s) + f(t+1,t), \quad t \in \mathbb{N}_a,$$

and

$$\Delta_t \sum_{s=t}^{b} g(t,s) = \sum_{s=t+1}^{b} \Delta_t g(t,s) - g(t,t), \quad t \in \mathbb{N}^b. \tag{1.12}$$

Proof We prove (1.12), being the other one completely analogous. We have

$$\begin{aligned}
\Delta_t \sum_{s=t}^{b} g(t,s) &= \sum_{s=t+1}^{b} g(t+1,s) - \sum_{s=t}^{b} g(t,s) \\
&= \sum_{s=t+1}^{b} g(t+1,s) - \sum_{s=t+1}^{b} g(t,s) - g(t,t) \\
&= \sum_{s=t+1}^{b} \Delta_t g(t,s) - g(t,t).
\end{aligned}$$

The proof is done.

1.2 Discrete Exponential Function

In this section we will describe the *discrete exponential function*, which plays a similar role in the discrete calculus as the exponential function $e^{\alpha t}$, $\alpha \in \mathbb{R}$, does in the continuous calculus.

We start by recalling that when α is a constant, then $y(t) = e^{\alpha t}$ is the unique solution of the initial value problem

$$y' = \alpha y, \quad y(0) = 1.$$

This is our motivation to define the discrete exponential function. Before we go into it, we need to introduce the following concept.

Definition 1.21 Consider $a \in \mathbb{R}$. The set of regressive functions is defined by

$$\mathcal{R} = \{p : \mathbb{N}_a \to \mathbb{R} : \ 1 + p(t) \neq 0 \text{ on } \mathbb{N}_a\}.$$

Now we suppose that $p \in \mathcal{R}$ and consider the first-order discrete initial value problem (IVP):

$$\Delta y(t) = p(t)y(t), \quad t \in \mathbb{N}_a, \tag{1.13}$$

$$y(s) = 1, \quad s \in \mathbb{N}_a. \tag{1.14}$$

In order to solve (1.13)–(1.14), let us introduce the following function:

$$\theta_p(t,s) = \begin{cases} \prod_{\tau=s}^{t-1}[1+p(\tau)] & \text{if } t \in \mathbb{N}_s, \\ \prod_{\tau=t}^{s-1}[1+p(\tau)]^{-1} & \text{if } t \in \mathbb{N}_a^{s-1}, \end{cases} \tag{1.15}$$

with the usual convention on products (which we follow hereafter), i.e., it is understood that for any function f

$$\prod_{\tau=s}^{s-1} f(\tau) = 1, \text{ for any } s \in \mathbb{N}_a.$$

Theorem 1.22 *Assume $p \in \mathcal{R}$ and $s \in \mathbb{N}_a$. Then, $\theta_p(t,s)$ introduced in* (1.15) *is the unique solution of the initial value problem* (1.13)–(1.14).

Proof We start by noticing that $\theta_p(s,s) = 1$, hence (1.14) is satisfied by $\theta_p(t,s)$. Moreover, equation (1.13) is equivalent to

$$y(t+1) = [1+p(t)]y(t), \quad t \in \mathbb{N}_a. \tag{1.16}$$

Now we assume that $t \in \mathbb{N}_s$. Then we get

$$\begin{aligned} y(s+1) &= [1+p(s)], \\ y(s+2) &= [1+p(s)][1+p(s+1)], \\ &\vdots \\ y(t) &= \prod_{\tau=s}^{t-1}[1+p(\tau)], \end{aligned}$$

which is precisely (1.15) for $t \in \mathbb{N}_s$. Next, assume $t \in \mathbb{N}_a^{s-1}$. Then, (1.16) is equivalent to

$$y(t) = \frac{1}{1+p(t)} y(t+1). \tag{1.17}$$

Hence, we get subsequently

$$\begin{aligned} y(s-1) &= \frac{1}{1+p(s-1)}, \\ y(s-2) &= \frac{1}{[1+p(s-2)][1+p(s-1)]}, \\ &\vdots \\ y(t) &= \prod_{\tau=t}^{s-1}[1+p(\tau)]^{-1}, \end{aligned}$$

which is (1.15) for $t \in \mathbb{N}_a^{s-1}$. The theorem is now proved.

Remark 1.23 It is easily seen that, for a regressive function p, the IVP

$$\Delta y(t) = p(t)y(t), \quad t \in \mathbb{N}_a,$$
$$y(s) = c, \quad s \in \mathbb{N}_a,\ c \in \mathbb{R},$$

has a unique solution.

Definition 1.24 The discrete exponential function corresponding to a function $p \in \mathcal{R}$ and based at $s \in \mathbb{N}_a$, which we denote by $e_p(t,s)$, is defined by

$$e_p(t,s) = \begin{cases} \prod_{\tau=s}^{t-1}[1+p(\tau)] & \text{if } t \in \mathbb{N}_s \\ \prod_{\tau=t}^{s-1}[1+p(\tau)]^{-1} & \text{if } t \in \mathbb{N}_a^{s-1}. \end{cases} \tag{1.18}$$

Remark 1.25 Note that the condition $p \in \mathcal{R}$ is necessary to define $e_p(t,s)$ as above only when $t \in \mathbb{N}_a^{s-1}$. So, from now on and without mentioning it further, when we consider $e_p(t,s) = \prod_{\tau=s}^{t-1}[1+p(\tau)]$, $t \in \mathbb{N}_s$, we will assume that $p : \mathbb{N}_a \to \mathbb{R}$ is a function.

Example 1.26 If $p(t) = p \in \mathbb{R}\backslash\{-1\}$, then the exponential function based at $s \in \mathbb{N}_a$ is (cf. (1.18))

$$e_p(t,s) = (1+p)^{t-s}, \quad t \in \mathbb{N}_a.$$

It is the solution of the IVP

$$\Delta y(t) = py(t), \quad t \in \mathbb{N}_a,$$
$$y(s) = 1, \quad s \in \mathbb{N}_a.$$

Remark 1.27 It is clear that the IVP

$$\Delta y(t) = p(t)y(t), \quad t \in \mathbb{N}_a,$$
$$y(a) = c, \quad c \in \mathbb{R},$$

has also a unique solution.

Next we address the nonhomogeneous linear equation.

Theorem 1.28 (Variation of Constants Formula) *Assume $p, q : \mathbb{N}_a \to \mathbb{R}$. Then, the initial value problem*

$$\Delta y(t) = p(t)y(t) + q(t), \quad t \in \mathbb{N}_a, \tag{1.19}$$
$$y(a) = A, \tag{1.20}$$

has a unique solution $\hat{y} : \mathbb{N}_a \to \mathbb{R}$, *which is given by*

$$\hat{y}(t) = Ae_p(t, a) + \sum_{s=a}^{t-1} e_p(t, s+1)q(s), \quad t \in \mathbb{N}_a.$$

Proof We start by showing that $\hat{y}$ satisfies the IVP (1.19)–(1.20). Using the Leibniz rule (cf. Proposition 1.20), we get

$$\begin{aligned}
\Delta\hat{y}(t) &= Ap(t)e_p(t, a) + \sum_{s=a}^{t-1} p(t)e_p(t, s+1)q(s) + e_p(t+1, t+1)q(t) \\
&= p(t)\left[Ae_p(t, a) + \sum_{s=a}^{t-1} e_p(t, s+1)q(s)\right] + q(t) \\
&= p(t)\hat{y}(t) + q(t),
\end{aligned}$$

for all $t \in \mathbb{N}_a$. Also $\hat{y}(a) = A$.

To prove uniqueness, let y_1, y_2 be two solutions of (1.19)–(1.20). Then the function $w(t) = (y_1 - y_2)(t)$ satisfies the IVP

$$\Delta w(t) = p(t)w(t), \quad t \in \mathbb{N}_a, \tag{1.21}$$

$$w(a) = 0. \tag{1.22}$$

By Remark 1.27, we conclude that $w(t) = 0$ on $\mathbb{N}_a$, i.e., $y_1(t) = y_2(t)$. The proof is done.

Example 1.29 ([41, Example 3.2.]) Suppose that we deposit \$2000 at the beginning of each year in an IRA that pays an annual interest rate of 8%. How much will we have in the IRA at the end of the tth year?

Let $y(t)$ be the amount of money in the IRA at the end of the tth year. Then,

$$\begin{aligned}
\Delta y(t) &= (y(t) + 2000) \cdot 0.08 + 2000, \quad y(0) = 0, \\
&= 0.08y(t) + 2160.
\end{aligned}$$

The solution y to the problem above is given by (cf. Theorem 1.28)

$$\begin{aligned}
y(t) &= \sum_{s=0}^{t-1} (1.08)^{t-s-1} 2160, \quad t \in \mathbb{N}_0 \\
&= 2160 \cdot (1.08)^{t-1} \sum_{s=0}^{t-1} \left(\frac{1}{1.08}\right)^s
\end{aligned}$$

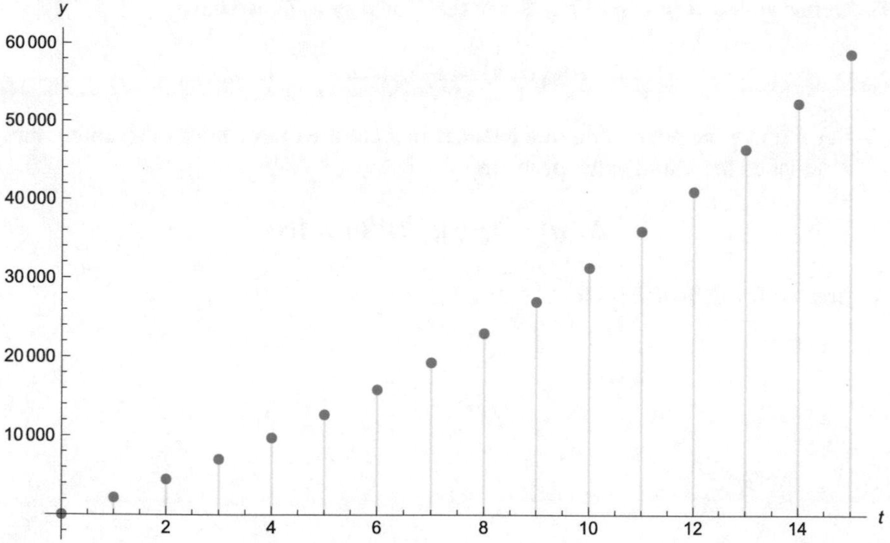

Fig. 1.1 $y(t)$ for $t \in \mathbb{N}_0^{15}$

$$= 2160 \cdot (1.08)^{t-1} \frac{\left(\frac{1}{1.08}\right)^t - 1}{\frac{1}{1.08} - 1}$$

$$= 27{,}000((1.08)^t - 1).$$

For example, at the end of 10 years we would have $y(10) \approx \$31{,}293$ (cf. in Fig. 1.1 the plot of $y(t)$ for $t \in \mathbb{N}_0^{15}$).

1.3 Exercises

1. Complete the proof of Theorem 1.3.
2. Show that

$$\Gamma(1/2) = \sqrt{\pi}. \tag{1.23}$$

3. Prove the equality in (1.4), whenever both sides are well-defined.
4. Complete the proof of Theorem 1.13.
5. Find:

   (a) $\sum_{k=1}^{n} k^3$.
   (b) $\sum_{s=0}^{t-1} s2^s$.

6. Define $p \oplus q = p + q + pq$. Show that, for $p, q \in \mathcal{R}$, we have

$$e_p(t, a)e_q(t, a) = e_{p\oplus q}(t, a), \quad t \in \mathbb{N}_a.$$

7. Let $P(t)$ be the population of a bacteria in a culture after t hours. Assuming that P satisfies the initial value problem

$$\Delta P(t) = 7P(t), \quad P(0) = 1000,$$

find the formula of $P(t)$ for $t \in \mathbb{N}_0$.

Chapter 2
Discrete Fractional Calculus

In this chapter we will develop the theory of discrete fractional calculus, using the operators introduced in the works by Miller and Ross [46] and Atici and Eloe [6].

As we shall see in this chapter, one of the peculiarities of the *forward* fractional difference is its domain shifting property, which makes, in a certain way, the study of this operator more complicated than its *backward* counterpart.[1]

2.1 Motivation

In this section we aim to provide the reader some reasoning in defining discrete fractional operators, i.e., the ν-th fractional sum and the ν-th fractional difference. Before going into that, we prove a (very) useful lemma.

Lemma 2.1 *Suppose that* $g : \mathbb{N}_a \times \mathbb{N}_a \to \mathbb{R}$. *Then, for* $b \in \mathbb{N}_a$, *we have*

$$\sum_{t=a}^{b-1} \sum_{s=a}^{t} g(t, s) = \sum_{s=a}^{b-1} \sum_{t=s}^{b-1} g(t, s).$$

Proof It is well-known that, for a function $\varphi : \mathbb{N}_a \to \mathbb{R}$, $\Delta\varphi(t) = 0$ on $\mathbb{N}_a$ if and only $\varphi(t) = c \in \mathbb{R}$ on $\mathbb{N}_a$.

[1] The reader may also find in the literature the nomenclatures "delta" and "nabla" operators for "forward" and "backward" operators, respectively.

R. A. C. Ferreira, *Discrete Fractional Calculus and Fractional Difference Equations*, SpringerBriefs in Mathematics,
https://doi.org/10.1007/978-3-030-92724-0_2

We define the function

$$\varphi(b) = \sum_{t=a}^{b-1}\sum_{s=a}^{t} g(t,s) - \sum_{s=a}^{b-1}\sum_{t=s}^{b-1} g(t,s), \quad b \in \mathbb{N}_a.$$

It is clear that $\varphi(a) = 0$. Moreover, using Proposition 1.20,

$$\begin{aligned}\Delta\varphi(b) &= \sum_{s=a}^{b} g(b,s) - \left(\sum_{s=a}^{b-1} g(b,s) + \sum_{t=b}^{b} g(t,b)\right)\\ &= 0.\end{aligned}$$

Therefore, $\varphi(b) = 0$, and the proof is done.

Remark 2.2 From Lemma 2.1, it is easy to deduce that

$$\sum_{t=a}^{b-1}\sum_{s=a}^{t-1} g(t,s) = \sum_{s=a}^{b-1}\sum_{t=s+1}^{b-1} g(t,s). \tag{2.1}$$

Indeed,

$$\begin{aligned}\sum_{t=a}^{b-1}\sum_{s=a}^{t-1} g(t,s) &= \sum_{t=a}^{b-1}\left(\sum_{s=a}^{t} g(t,s) - g(t,t)\right)\\ &= \sum_{s=a}^{b-1}\sum_{t=s}^{b-1} g(t,s) - \sum_{s=a}^{b-1} g(s,s)\\ &= \sum_{s=a}^{b-1}\sum_{t=s+1}^{b-1} g(t,s).\end{aligned}$$

In order to motivate the definition of the ν-th fractional sum, let us start by calculating the n-fold summation of a function $f : \mathbb{N}_a \to \mathbb{R}$.

- $n = 1$: We have $\sum_{s=a}^{t-1} f(s)$.
- $n = 2$: We have

$$\sum_{s=a}^{t-1}\sum_{r=a}^{s-1} f(r) = \sum_{r=a}^{t-1}\sum_{s=r+1}^{t-1} f(r) = \sum_{r=a}^{t-1}(t-(r+1))f(r).$$

- $n = 3$: We have (recall Example 1.16)

$$\sum_{s=a}^{t-1}\sum_{r=a}^{s-1}(s-(r+1))f(r) = \sum_{r=a}^{t-1}\sum_{s=r+1}^{t-1}(s-(r+1))f(r) = \sum_{r=a}^{t-1}\frac{(t-(r+1))^{\underline{2}}}{2}f(r).$$

Proceeding as above, we may show that

$$\sum_{s_1=a}^{t-1}\sum_{s_2=a}^{s_1-1}\cdots\sum_{s_n=a}^{s_{n-1}-1} f(s_n)=\sum_{s=a}^{t-1}\frac{(t-(s+1))^{\underline{n-1}}}{(n-1)!}f(s), \quad n\in\mathbb{N}_1.$$

By noticing that

$$(t-(s+1))^{\underline{n-1}}=0,\ \ s=t-1,t-2,t-n+1,$$

we define

$$\Delta_a^{-n}f(t)=\frac{1}{\Gamma(n)}\sum_{s=a}^{t-n}(t-(s+1))^{\underline{n-1}}f(s), \quad t\in\mathbb{N}_{a+n}. \tag{2.2}$$

We close this section by observing that the right-hand side of (2.2) is meaningful for any real number $n>0$. This is the key point into defining the ν-th fractional sum ($\nu>0$), and subsequently the ν-th fractional difference, in the next section.

2.2 The Fractional Sums and Differences

The previous section indicated us a way of introducing the concept of fractional sum. We actually introduce here the concepts of left and right fractional sums, respectively (the reason will be clear to the reader when we develop the theory of the *calculus of variations* in Chap. 4).

Definition 2.3 Consider $a,b\in\mathbb{R}$, and let $f:\mathbb{N}_a\to\mathbb{R}$ and $g:\mathbb{N}^b\to\mathbb{R}$. Then we define the *left fractional sum of* f with order $\nu>0$ to be the function $\Delta_a^{-\nu}f:\mathbb{N}_{a+\nu-1}\to\mathbb{R}$ given by

$$\Delta_a^{-\nu}f(t)=\begin{cases}\frac{1}{\Gamma(\nu)}\sum_{s=a}^{t-\nu}(t-(s+1))^{\underline{\nu-1}}f(s) & \text{if } t\in\mathbb{N}_{a+\nu},\\ 0 \quad \text{if } t=a+\nu-1.\end{cases}$$

The *right fractional sum of* g of order $\nu>0$ is the function ${}_b\Delta^{-\nu}g:\mathbb{N}^{b+1-\nu}\to\mathbb{R}$ defined by

$${}_b\Delta^{-\nu}g(t)=\begin{cases}\frac{1}{\Gamma(\nu)}\sum_{s=t+\nu}^{b}(s-(t+1))^{\underline{\nu-1}}g(s), & \text{if } t\in\mathbb{N}^{b-\nu},\\ 0, \ \text{if } t=b+1-\nu.\end{cases}$$

Remark 2.4 We would like that if one took $\nu=0$ in any of the fractional sums above, then we would get the identity operator. Let us gain some intuition of how the left fractional sum above[2] behaves near $\nu=0$.

[2] The analysis for the right fractional sum is completely analogous.

We have, for $\nu > 0$ and $t \in \mathbb{N}_{a+\nu}$,

$$\frac{1}{\Gamma(\nu)}\sum_{s=a}^{t-\nu}(t-(s+1))^{\underline{\nu-1}}f(s) = f(t-\nu) + \sum_{s=a}^{t-\nu-1}\frac{\Gamma(t-s)}{\Gamma(\nu)\Gamma(t-s-\nu+1)}f(s)$$
$$= f(t-\nu) + \sum_{s=a+\nu}^{t-1}\frac{\Gamma(t-s+\nu)}{\Gamma(\nu)\Gamma(t-s+1)}f(s-\nu)$$
$$= f(t-\nu) + \sum_{s=a+\nu}^{t-1}\frac{\prod_{i=0}^{t-s-1}(\nu+i)}{\Gamma(t-s+1)}f(s-\nu).$$

Formally, if we let $\nu = 0$ in the right-hand side of the previous equality, we arrive at $f(t)$.

We, therefore, extend Definition 2.3 to $\nu = 0$ by letting

$$\Delta_a^{-0}f(t) = f(t),\ t \in \mathbb{N}_a \text{ and } {}_b\Delta^{-0}g(t) = g(t),\ t \in \mathbb{N}^b.$$

Example 2.5 Let us calculate the right fractional sum of the constant function $g(t) = c$, with $c \in \mathbb{R}$. Let $\nu > 0$. We have

$${}_b\Delta^{-\nu}g(t) = \frac{c}{\Gamma(\nu)}\sum_{s=t+\nu}^{b}(s-(t+1))^{\underline{\nu-1}} = c\frac{(b-t)^{\underline{\nu}}}{\Gamma(\nu+1)}, \quad t \in \mathbb{N}^{b-\nu}.$$

Definition 2.6 Consider $a, b \in \mathbb{R}$ and let $f : \mathbb{N}_a \to \mathbb{R}$ and $g : \mathbb{N}^b \to \mathbb{R}$. The *Riemann–Liouville left fractional difference of* f and the *Riemann–Liouville right fractional difference of* g of order $0 < \alpha \leq 1$ are defined, respectively, as

$$\Delta_a^{\alpha}f(t) = \Delta[\Delta_a^{-(1-\alpha)}f](t), \quad t \in \mathbb{N}_{a+1-\alpha}, \tag{2.3}$$

and

$${}_b\Delta^{\alpha}g(t) = -\Delta[{}_b\Delta^{-(1-\alpha)}g](t), \quad t \in \mathbb{N}^{b-2+\alpha}.$$

The *Caputo left fractional difference of* f and the *Caputo right fractional difference of* g of order $0 < \alpha \leq 1$ are defined, respectively, as

$${}_*\Delta_a^{\alpha}f(t) = \Delta_a^{-(1-\alpha)}[\Delta f](t), \quad t \in \mathbb{N}_{a+1-\alpha},$$

and

$${}_b\Delta_*^{\alpha}g(t) = -{}_{b-1}\Delta^{-(1-\alpha)}[\Delta g](t), \quad t \in \mathbb{N}^{b-2+\alpha}.$$

Remark 2.7 Observe that when $\alpha = 1$ we get the usual forward difference operator (Δ) for the left fractional differences and its symmetric $(-\Delta)$ for the right fractional differences.

Remark 2.8 Observe that all the fractional sums and differences defined above are linear operators.

Remark 2.9 We can define higher order fractional differences as in, e.g., [35, Section 2]. However, since in this work we only deal with differences of order $0 < \alpha < 1$, we opted to write Definition 2.6 as above.

Remark 2.10 Note that, upon using Proposition 1.20, we may write Δ_a^α, with $0 < \alpha < 1$, as

$$\begin{aligned}\Delta_a^\alpha f(t) &= \frac{\Delta}{\Gamma(1-\alpha)} \sum_{s=a-\alpha}^{t-1} (t-(s+\alpha+1))^{\underline{-\alpha}} f(s+\alpha)\\ &= \frac{1}{\Gamma(-\alpha)} \sum_{s=a-\alpha}^{t-1} (t-(s+\alpha+1))^{\underline{-\alpha-1}} f(s+\alpha) + f(t+\alpha)\\ &= \frac{1}{\Gamma(-\alpha)} \sum_{s=a}^{t+\alpha} (t-(s+1))^{\underline{-\alpha-1}} f(s),\end{aligned}$$

for $t \in \mathbb{N}_{a+1-\alpha}$.

Remark 2.11 Miller and Ross [46] defined a fractional sum–difference operator of order $\nu \in \mathbb{R}\backslash\mathbb{N}^0$ by

$$^{M-R}\Delta_a{}^{\nu} f(t) = \frac{1}{\Gamma(\nu)} \sum_{s=a}^{t-\nu} (t-(s+1))^{\underline{\nu-1}} f(s), \quad t \in \mathbb{N}_{a+\nu}.$$

The definitions of fractional sum and fractional difference (of left Riemann–Liouville type) as in Definitions 2.3 and 2.6 should be attributed to Atici and Eloe [6], though they did not set a value for $\Delta_a^{-\nu} f(a+\nu-1)$, $\nu > 0$. That precluded them to include the point $t = a - \alpha$ in the definition given as in (2.3). Let us mention the differences between the fractional differences introduced by Miller and Ross and by Atici and Eloe (since both definitions coincide for the fractional sum, i.e., when $\nu > 0$).

To start with observe that the Miller–Ross fractional operator is not formally defined for when the power is a negative integer. So, consider $0 < \mu < 1$. Then, the Miller–Ross fractional difference of order μ is

$$^{M-R}\Delta_a^{\mu} f(t) = \frac{1}{\Gamma(-\mu)} \sum_{s=a}^{t+\mu} (t-(s+1))^{\underline{-\mu-1}} f(s), \quad t \in \mathbb{N}_{a-\mu}, \tag{2.4}$$

while the Atici–Eloe one is given by (cf. Remark 2.10)

$$ {}^{A-E}\Delta_a^{\mu} f(t) = \frac{1}{\Gamma(-\mu)} \sum_{s=a}^{t+\mu} (t-(s+1))^{\underline{-\mu-1}} f(s), \quad t \in \mathbb{N}_{a+1-\mu}. \tag{2.5} $$

Therefore, the Miller–Ross fractional difference and the Atici–Eloe fractional difference coincide but have different domains of definition. Nevertheless, one may observe that, by setting $\Delta_a^{-\nu} f(a+\nu-1) = 0$ as we did in Definition 2.3 then, with $0 < \alpha < 1$, $\Delta_a^{\alpha} f(t)$ at $t = a - \alpha$ in (2.3) is just $f(a)$ and (2.4) agrees with (2.5) at each point of $\mathbb{N}_{a-\mu}$.

The Riemann–Liouville and the Caputo fractional differences are related by the content of the following result.

Proposition 2.12 *Consider $a, b \in \mathbb{R}$, and let $f : \mathbb{N}_a \to \mathbb{R}$ and $g : \mathbb{N}^b \to \mathbb{R}$. For $0 < \alpha < 1$, we have*

$$ {}_*\Delta_a^{\alpha} f(t) = \Delta_a^{\alpha} f(t) - \frac{(t-a)^{\underline{-\alpha}}}{\Gamma(1-\alpha)} f(a), \quad t \in \mathbb{N}_{a+1-\alpha}, \tag{2.6} $$

and

$$ {}_b\Delta_*^{\alpha} g(t) = {}_b\Delta^{\alpha} g(t) - \frac{(b-(t+1))^{\underline{-\alpha}}}{\Gamma(1-\alpha)} g(b), \quad t \in \mathbb{N}^{b-2+\alpha}. $$

Proof We will prove the formula for the right fractional differences, leaving the other one for the reader. Using (1.11), we get

$$ \begin{aligned} {}_b\Delta_*^{\alpha} g(t) &= -\frac{1}{\Gamma(1-\alpha)} \sum_{s=t+1-\alpha}^{b-1} (s-(t+1))^{\underline{-\alpha}} \Delta g(s) \\ &= -\frac{1}{\Gamma(1-\alpha)} \left(\left[(s-(t+1)^{\underline{-\alpha}}) g(s) \right]_{s=t+1-\alpha}^{s=b} \right. \\ &\qquad \left. - \sum_{s=t+1-\alpha}^{b-1} (-\alpha)(s-(t+1))^{\underline{-\alpha-1}} g(s+1) \right) (*), \end{aligned} $$

and after some simplifications

$$(*) = -\frac{(b-(t+1))^{\underline{-\alpha}} g(b)}{\Gamma(1-\alpha)} - \frac{1}{\Gamma(1-\alpha)} \sum_{s=t+1}^{b+\alpha-1} \Delta_t (s-\alpha-t)^{\underline{-\alpha}} g(s+1-\alpha) + g(t+1-\alpha).$$

Now we apply (1.12) to conclude that

$$\begin{aligned}
(*) &= -\frac{(b-(t+1))^{\underline{-\alpha}} g(b)}{\Gamma(1-\alpha)} - \Delta_t \frac{1}{\Gamma(1-\alpha)} \sum_{s=t}^{b+\alpha-1} (s-\alpha-t)^{\underline{-\alpha}} g(s+1-\alpha) \\
&= -\frac{(b-(t+1))^{\underline{-\alpha}} g(b)}{\Gamma(1-\alpha)} - \Delta_t \frac{1}{\Gamma(1-\alpha)} \sum_{s=t+1-\alpha}^{b} (s-(t+1))^{\underline{-\alpha}} g(s) \\
&= -\frac{(b-(t+1))^{\underline{-\alpha}} g(b)}{\Gamma(1-\alpha)} + {}_b\Delta^{\alpha} g(t),
\end{aligned}$$

and the proof is done.

Remark 2.13 It is clear from the previous result that if $0 < \alpha < 1$, then the left (right, respectively) Riemann–Liouville and Caputo fractional differences coincide if and only if $f(a) = 0$ (respectively, $g(b) = 0$).

The following result is of independent interest but will be, nevertheless, very useful within the calculus of variations theory, which we will develop later in Chap. 4.

Theorem 2.14 (Fractional Summation by Parts) *Let* $0 < \alpha \leq 1$. *For functions* $f : \mathbb{N}_a^{b-1} \to \mathbb{R}$ *and* $g : \mathbb{N}_{a+\alpha-1}^{b+\alpha-1} \to \mathbb{R}$, *the following equality holds:*

$$\begin{aligned}
&\sum_{t=a}^{b-1} f(t)_* \Delta_{a+\alpha-1}^{\alpha} g(t) \\
&\quad = g(b+\alpha-1) f(b-1) - g(a+\alpha-1)_{b-1}\Delta^{-(1-\alpha)} f(a+\alpha-1) \qquad (2.7) \\
&\quad + \sum_{t=a}^{b-2} g(t+\alpha)_{b-1}\Delta^{\alpha} f(t+\alpha-1).
\end{aligned}$$

Proof The case $\alpha = 1$ follows immediately from (1.11). For $0 < \alpha < 1$, we have

$$\begin{aligned}
&\sum_{t=a}^{b-1} f(t)_* \Delta^{\alpha}_{a+\alpha-1} g(t)\\
&= \sum_{t=a}^{b-1} f(t) \Delta^{-(1-\alpha)}_{a+\alpha-1} \Delta g(t)\\
&= \sum_{t=a}^{b-1} f(t) \frac{1}{\Gamma(1-\alpha)} \sum_{s=a}^{t} (t-(s+\alpha-1+1))^{\underline{-\alpha}} \Delta g(s+\alpha-1)\\
&= \sum_{t=a}^{b-1} \Delta g(t+\alpha-1) \frac{1}{\Gamma(1-\alpha)} \sum_{s=t}^{b-1} (s-(t+\alpha-1+1))^{\underline{-\alpha}} f(s)\\
&= \Delta g(b+\alpha-2) f(b-1)\\
&\quad + \sum_{t=a}^{b-2} \Delta g(t+\alpha-1) \frac{1}{\Gamma(1-\alpha)} \sum_{s=t}^{b-1} (s-(t+\alpha-1+1))^{\underline{-\alpha}} f(s)\\
&= \Delta g(b+\alpha-2) f(b-1) + \sum_{t=a}^{b-2} \Delta g(t+\alpha-1)_{b-1} \Delta^{-(1-\alpha)} f(t+\alpha-1)\\
&= \Delta g(b+\alpha-2) f(b-1) + \Big[g(t+\alpha-1)_{b-1} \Delta^{-(1-\alpha)} f(t+\alpha-1) \Big]_{t=a}^{t=b-1}\\
&\quad + \sum_{t=a}^{b-2} g(t+\alpha)_{b-1} \Delta^{\alpha} f(t+\alpha-1),
\end{aligned}$$

where we have used Lemma 2.1 and equality (1.11). The proof is done.

2.3 Power Rules

We start by proving a binomial-type formula, needed in the sequel.

Lemma 2.15 (Discrete Analogue of the Binomial Theorem) *Let $x, y \in \mathbb{R}$ and $n \in \mathbb{N}_0$. Then,*

$$(x+y)_n = \sum_{k=0}^{n} \binom{n}{k} (x)_{n-k} (y)_k. \tag{2.8}$$

Proof We use induction on n. The case $n = 0$ is obvious. Now, suppose that (2.8) holds for any real numbers x, y. Then,

$$\begin{aligned}
&\sum_{k=0}^{n+1}\binom{n+1}{k}(x)_{n+1-k}(y)_k \\
&= (x)_{n+1} + (y)_{n+1} + \sum_{k=1}^{n}\left[\binom{n}{k}+\binom{n}{k-1}\right](x)_{n+1-k}(y)_k \\
&= \sum_{k=0}^{n}\binom{n}{k}(x)_{n+1-k}(y)_k + \sum_{k=0}^{n-1}\binom{n}{k}(x)_{n-k}(y)_{k+1} + (y)_{n+1} \\
&= \sum_{k=0}^{n}\binom{n}{k}(x)_{n-k}(x+n-k)(y)_k + \sum_{k=0}^{n}\binom{n}{k}(x)_{n-k}(y)_k(y+k) \\
&= (x+y+n)(x+y)_n = (x+y)_{n+1},
\end{aligned}$$

and the proof is done.

Remark 2.16 It can also be shown that

$$(x+y)^{\underline{n}} = \sum_{k=0}^{n}\binom{n}{k}x^{\underline{n-k}}y^{\underline{k}},$$

for all $x, y \in \mathbb{R}$ and $n \in \mathbb{N}_0$.

It follows the *fractional sum power rule*.

Theorem 2.17 *Let $a \in \mathbb{R}$. Assume $\mu \notin \mathbb{N}^{-1}$ and $\nu > 0$. Then,*

$$\Delta_{a+\mu}^{-\nu}[(s-a)^{\underline{\mu}}](t) = \frac{\Gamma(\mu+1)}{(t-a-\mu-\nu)!}(\mu+\nu+1)_{t-a-\mu-\nu}, \text{ for } t \in \mathbb{N}_{a+\mu+\nu}. \tag{2.9}$$

Proof Let $t \in \mathbb{N}_{a+\mu+\nu}$. Then,

$$\begin{aligned}
\Delta_{a+\mu}^{-\nu}[(s-a)^{\underline{\mu}}](t) &= \frac{1}{\Gamma(\nu)}\sum_{s=a+\mu}^{t-\nu}(t-(s+1))^{\underline{\nu-1}}(s-a)^{\underline{\mu}} \\
&= \sum_{s=a+\mu}^{t-\nu}\frac{1}{\Gamma(\nu)}\frac{\Gamma(t-s)}{\Gamma(t-s+1-\nu)}\frac{\Gamma(s-a+1)}{\Gamma(s-a+1-\mu)} \\
&= \sum_{s=0}^{t-a-\mu-\nu}\frac{1}{\Gamma(\nu)}\frac{\Gamma(t-s-a-\mu)}{\Gamma(t-s-a-\mu+1-\nu)}\frac{\Gamma(s+\mu+1)}{\Gamma(s+1)}.
\end{aligned}$$

Put $n = t - a - \mu - \nu \in \mathbb{N}_0$. Then we get from the previous equality

$$\begin{aligned}
\Delta_{a+\mu}^{-\nu}[(s-a)^{\underline{\mu}}](t) &= \sum_{s=0}^{n} \frac{1}{\Gamma(\nu)} \frac{\Gamma(n-s+\nu)}{\Gamma(n-s+1)} \frac{\Gamma(s+\mu+1)}{\Gamma(s+1)} \\
&= \frac{\Gamma(\mu+1)}{n!} \sum_{s=0}^{n} \frac{n!}{(n-s)!s!} \frac{\Gamma(n-s+\nu)}{\Gamma(\nu)} \frac{\Gamma(s+\mu+1)}{\Gamma(\mu+1)} \\
&= \frac{\Gamma(\mu+1)}{n!} \sum_{s=0}^{n} \binom{n}{s} (\nu)_{n-s} (\mu+1)_s \\
&= \frac{\Gamma(\mu+1)}{n!} (\mu+\nu+1)_n,
\end{aligned}$$

where we have used (2.8). Finally, substituting n by $t - a - \mu - \nu$, we obtain (2.9).

Corollary 2.18 *Let $a \in \mathbb{R}$. Assume $\mu \notin \mathbb{N}^{-1}$ and $\nu > 0$.*
If $\mu + \nu \notin \mathbb{N}^{-1}$, then

$$\Delta_{a+\mu}^{-\nu}[(s-a)^{\underline{\mu}}](t) = \frac{\Gamma(\mu+1)}{\Gamma(\mu+\nu+1)} (t-a)^{\underline{\mu+\nu}}, \text{ for } t \in \mathbb{N}_{a+\mu+\nu}, \tag{2.10}$$

while if $\mu + \nu \in \mathbb{N}^{-1}$, then

$$\Delta_{a+\mu}^{-\nu}[(s-a)^{\underline{\mu}}](t) = 0, \text{ for } t \in \mathbb{N}_a. \tag{2.11}$$

Proof If $\mu + \nu \notin \mathbb{N}^{-1}$, then

$$(\mu+\nu+1)_{t-a-\mu-\nu} = \frac{\Gamma(t-a+1)}{\Gamma(\mu+\nu+1)}, \quad t \in \mathbb{N}_{a+\mu+\nu}.$$

Hence, (2.9) becomes

$$\Delta_{a+\mu}^{-\nu}[(s-a)^{\underline{\mu}}](t) = \frac{\Gamma(\mu+1)}{\Gamma(\mu+\nu+1)} \frac{\Gamma(t-a+1)}{\Gamma(t-a+1-\mu-\nu)},$$

which is just (2.10).

If $\mu + \nu \in \mathbb{N}^{-1}$, then, using Definition 1.8, $(\mu+\nu+1)_{t-a-\mu-\nu} = 0$ for $t \in \mathbb{N}_a$, and (2.11) follows.

Corollary 2.19 *Suppose that $\nu > 0$ and $\mu + \nu \in \mathbb{N}^{-1}$ with $\mu \notin \mathbb{N}^{-1}$. Then,*

$$\sum_{k=0}^{n} \binom{n}{k} \Gamma(n+\nu-k)\Gamma(k+\mu+1) = 0, \ n \in \mathbb{N}_{-(\mu+\nu)}.$$

Proof Consider $a = 0$. Then, we have by (2.11)

$$\frac{1}{\Gamma(\nu)}\sum_{k=\mu}^{t-\nu}\frac{\Gamma(t-k)}{\Gamma(t-k+1-\nu)}\frac{\Gamma(k+1)}{\Gamma(k+1-\mu)}=0, \quad t\in\mathbb{N}_0,$$

which is equivalent to

$$\sum_{k=0}^{t-(\mu+\nu)}\frac{\Gamma(t-(k+\mu))}{\Gamma(t-(k+\mu)+1-\nu)}\frac{\Gamma(k+\mu+1)}{\Gamma(k+\mu+1-\mu)}=0.$$

Let $n = t-(\mu+\nu)$ and note that $n \in \mathbb{N}_{-(\mu+\nu)}$. It follows from the previous equality that

$$\sum_{k=0}^{n}\frac{\Gamma(n+\nu-k)}{(n-k)!}\frac{\Gamma(k+\mu+1)}{k!}=0,$$

from which the result is an immediate consequence.

We now turn to consider power rules but using fractional difference operators.

Theorem 2.20 (Fractional Difference Power Rules) *Let $a \in \mathbb{R}$. Assume $0 < \alpha \leq 1$, $\mu \notin \mathbb{N}^0$, and $\mu - \alpha + 1 \notin \mathbb{N}^0$. Then,*

$$\Delta^{\alpha}_{a+\mu-1}[(s-a)^{\underline{\mu}}](t)=\frac{\Gamma(\mu+1)}{\Gamma(\mu-\alpha+1)}(t-a)^{\underline{\mu-\alpha}}, \text{ for } t\in\mathbb{N}_{a+\mu-\alpha}, \tag{2.12}$$

and

$$ {}_*\Delta^{\alpha}_{a+\mu-1}[(s-a)^{\underline{\mu}}](t)=\frac{\Gamma(\mu+1)}{\Gamma(\mu-\alpha+1)}(t-a)^{\underline{\mu-\alpha}}, \text{ for } t\in\mathbb{N}_{a+\mu-\alpha}. \tag{2.13}$$

Proof By Corollary 2.18, we have

$$\Delta^{-(1-\alpha)}_{a+\mu-1}[(s-a)^{\underline{\mu}}](t)=\Delta^{-(1-\alpha)}_{a+\mu}[(s-a)^{\underline{\mu}}](t)=\frac{\Gamma(\mu+1)}{\Gamma(\mu+2-\alpha)}(t-a)^{\underline{\mu+1-\alpha}},$$

for all $t \in \mathbb{N}_{a+\mu+1-\alpha}$. By the definition of the falling function, clearly the previous equality also holds for $t = a + \mu - \alpha$. So, apply the Δ operator to both sides of the equality and use (1.2) to obtain (2.12).

In order to obtain (2.13) just note that

$$ {}_*\Delta^{\alpha}_{a+\mu-1}[(s-a)^{\underline{\mu}}](t)=\mu\Delta^{-(1-\alpha)}_{a+\mu-1}[(s-a)^{\underline{\mu-1}}](t)=\frac{\Gamma(\mu+1)}{\Gamma(\mu-\alpha+1)}(t-a)^{\underline{\mu-\alpha}}.$$

The proof is done.

We may deduce from the previous results analogous ones for the right fractional operators. We formalize this for the right fractional sum.

Theorem 2.21 *Let $b \in \mathbb{R}$. Assume $\mu \notin \mathbb{N}^{-1}$ and $\nu > 0$. Then,*

$$ {}_{b-\mu}\Delta^{-\nu}[(b-s)^{\underline{\mu}}](t) = \frac{\Gamma(\mu+1)}{(b-t-\mu-\nu)!}(\mu+\nu+1)_{b-t-\mu-\nu}, \ \textit{for } t \in \mathbb{N}^{b-\mu-\nu}. $$

Moreover,

$$ {}_{b-\mu}\Delta^{-\nu}[(b-s)^{\underline{\mu}}](t) = \frac{\Gamma(\mu+1)}{\Gamma(\mu+\nu+1)}(b-t)^{\underline{\mu+\nu}}, \ \textit{for } t \in \mathbb{N}^{b-\mu-\nu} \textit{ and } \mu+\nu \notin \mathbb{N}^{-1}, $$

and

$$ {}_{b-\mu}\Delta^{-\nu}[(b-s)^{\underline{\mu}}](t) = 0, \ \textit{for } t \in \mathbb{N}^{b} \textit{ and } \mu+\nu \in \mathbb{N}^{-1}. $$

Proof Consider $\mu \neq 0$ (the case $\mu = 0$ is considered in Example (2.5)). We have

$$ \begin{aligned} &{}_{b-\mu}\Delta^{-\nu}[(b-s)^{\underline{\mu}}](t) \\ &= \frac{1}{\Gamma(\nu)} \sum_{s=t+\nu}^{b-\mu} (s-(t+1))^{\underline{\nu-1}}(b-s)^{\underline{\mu}} \\ &= \frac{1}{\Gamma(\nu)} \left(\left[(b-s)^{\underline{\mu}} \frac{(s-(t+1))^{\underline{\nu}}}{\nu} \right]_{s=t+\nu}^{s=b-\mu+1} \right. \\ &\qquad \left. + \frac{\mu}{\nu} \sum_{s=t+\nu}^{b-\mu} (b-(s+1))^{\underline{\mu-1}}(s-t)^{\underline{\nu}} \right) \\ &= \frac{\mu}{\nu}\frac{\Gamma(\mu)}{\Gamma(\nu)}\frac{1}{\Gamma(\mu)} \sum_{s=\nu}^{(b-t)-\mu} (b-t-(s+1))^{\underline{\mu-1}} s^{\underline{\nu}} \\ &= \frac{\Gamma(\mu+1)}{\Gamma(\nu+1)} \Delta_{\nu}^{-\mu}[s^{\underline{\nu}}](b-t), \end{aligned} $$

where we have used the summation by parts formula (1.11). The result now follows from Theorem 2.17 and Corollary 2.18.

Example 2.22 Find

$$ {}_{*}\Delta_{\frac{3}{2}}^{\frac{1}{2}}[(s-1)^{\underline{\frac{3}{2}}}](t), \quad t \in \mathbb{N}_2. $$

We have (cf. (1.23))

$$
{}_*\Delta_{\frac{3}{2}}^{\frac{1}{2}}[(s-1)^{\underline{\frac{3}{2}}}](t) = \frac{\Gamma(\frac{3}{2}+1)}{\Gamma(\frac{3}{2}-\frac{1}{2}+1)}(t-1)^{\underline{\frac{3}{2}-\frac{1}{2}}} = \frac{3}{2}\Gamma\left(\frac{1}{2}+1\right)(t-1)
$$

$$
= \frac{3}{4}\Gamma\left(\frac{1}{2}\right)(t-1) = \frac{3\sqrt{\pi}}{4}(t-1),
$$

for $t \in \mathbb{N}_2$.

2.4 Composition Rules

We start with the composition for fractional sums.

Theorem 2.23 *Let $a, b \in \mathbb{R}$. Assume $\mu, \nu > 0$.*
If $f : \mathbb{N}_a \to \mathbb{R}$, then

$$
\Delta_{a+\nu}^{-\mu}[\Delta_a^{-\nu} f](t) = \Delta_a^{-(\mu+\nu)} f(t), \quad t \in \mathbb{N}_{a+\mu+\nu}. \tag{2.14}
$$

If $g : \mathbb{N}^b \to \mathbb{R}$, then

$$
{}_{b-\nu}\Delta^{-\mu}[{}_b\Delta^{-\nu} g](t) = {}_b\Delta^{-(\mu+\nu)} g(t), \quad t \in \mathbb{N}^{b-\mu-\nu}. \tag{2.15}
$$

Proof We will only prove the formula for the left fractional sum, leaving the other one to the reader.

Assume $t \in \mathbb{N}_{a+\mu+\nu}$. Then,

$$
\Delta_{a+\nu}^{-\mu}[\Delta_a^{-\nu} f](t)
$$

$$
= \frac{1}{\Gamma(\mu)\Gamma(\nu)} \sum_{s=a+\nu}^{t-\mu} (t-(s+1))^{\underline{\mu-1}} \sum_{k=a}^{s-\nu} (s-(k+1))^{\underline{\nu-1}} f(k)
$$

$$
= \frac{1}{\Gamma(\mu)\Gamma(\nu)} \sum_{s=a}^{t-\mu-\nu} \sum_{k=a}^{s} (t-(s+\nu+1))^{\underline{\mu-1}} (s+\nu-(k+1))^{\underline{\nu-1}} f(k)
$$

$$
= \frac{1}{\Gamma(\mu)\Gamma(\nu)} \sum_{k=a}^{t-(\mu+\nu)} \sum_{s=k}^{t-(\mu+\nu)} (t-(s+\nu+1))^{\underline{\mu-1}} (s+\nu-(k+1))^{\underline{\nu-1}} f(k),
$$

where we have used Lemma 2.1 to get the last equality. Proceeding, we obtain from the last expression

$$\frac{1}{\Gamma(\mu)\Gamma(\nu)}\sum_{k=a}^{t-(\mu+\nu)}\sum_{s=k}^{t-(\mu+\nu)}(t-(s+\nu+1))^{\underline{\mu-1}}(s+\nu-(k+1))^{\underline{\nu-1}}f(k)$$

$$=\frac{1}{\Gamma(\nu)}\sum_{k=a}^{t-(\mu+\nu)}\frac{1}{\Gamma(\mu)}\sum_{s=k+\nu-1}^{t-1-\mu}(t-1-(s+1))^{\underline{\mu-1}}(s-k)^{\underline{\nu-1}}f(k)$$

$$=\frac{1}{\Gamma(\nu)}\sum_{k=a}^{t-(\mu+\nu)}\underbrace{\frac{1}{\Gamma(\mu)}\sum_{s=k+\nu-1}^{t-1-\mu}(t-1-(s+1))^{\underline{\mu-1}}(s-k)^{\underline{\nu-1}}}_{\Delta_{k+\nu-1}^{-\mu}[(s-k)^{\underline{\nu-1}}](t-1)}f(k)$$

$$=\frac{1}{\Gamma(\mu+\nu)}\sum_{k=a}^{t-(\mu+\nu)}(t-(k+1))^{\underline{\mu+\nu-1}}f(k)=\Delta_a^{-(\mu+\nu)}f(t),$$

where, in virtue of $\mu+\nu-1\notin\mathbb{N}^{-1}$, we have used (2.10). The proof is done.

Theorem 2.24 *Assume $f:\mathbb{N}_a\to\mathbb{R}$ and $g:\mathbb{N}^b\to\mathbb{R}$. Moreover, let $0<\alpha\le 1$. Then,*

$${}_*\Delta_{a+\alpha-1}^{\alpha}\Delta_a^{-\alpha}f(t)=f(t),\quad t\in\mathbb{N}_a,$$

and

$${}_{b-\alpha}\Delta^{\alpha}{}_b\Delta^{-\alpha}g(t)=g(t+1),\quad t\in\mathbb{N}^{b-1}.$$

Proof We start by noticing that using (2.14), we get

$$\Delta_{a+\alpha-1}^{\alpha}\Delta_a^{-\alpha}f(t)=\Delta\Delta_{a+\alpha-1}^{-(1-\alpha)}\Delta_a^{-\alpha}f(t)=\Delta\Delta_{a+\alpha}^{-(1-\alpha)}\Delta_a^{-\alpha}f(t)=f(t),$$

for all $t\in\mathbb{N}_a$, where we have also used the fundamental theorem of calculus. Now, since $\Delta_a^{-\alpha}f(a+\alpha-1)=0$, we have by Proposition 2.12 ${}_*\Delta_{a+\alpha-1}^{\alpha}\Delta_a^{-\alpha}f(t)=\Delta_{a+\alpha-1}^{\alpha}\Delta_a^{-\alpha}f(t)$, from which the result follows.

For completeness, we also prove the formula for the right fractional operator. We have

$${}_{b-\alpha}\Delta^{\alpha}{}_b\Delta^{-\alpha}g(t)=-\Delta[{}_{b-\alpha}\Delta^{-(1-\alpha)}{}_b\Delta^{-\alpha}g](t)=-\Delta[{}_b\Delta^{-1}g](t),$$

upon using Theorem 2.23. Now,

$$-\Delta[{}_b\Delta^{-1}g](t) = -\left(\sum_{s=t+2}^{b} g(s) - \sum_{s=t+1}^{b} g(s)\right) = g(t+1),$$

and the proof is done.

Before we proceed, let us provide a generalization of the previous theorem.

Theorem 2.25 *Assume* $f : \mathbb{N}_a \to \mathbb{R}$. *Moreover, let* $0 < \alpha \le 1$ *and* $\beta \ge 0$ *be such that* $0 \le \alpha - \beta \le 1$. *Then,*

$${}_*\Delta^{\alpha}_{a+\beta-1}\Delta_a^{-\beta} f(t) = \Delta_a^{\alpha-\beta} f(t), \quad t \in \mathbb{N}_{a+\beta-\alpha}.$$

Proof Using (2.14), we get

$$\Delta^{\alpha}_{a+\beta-1}\Delta_a^{-\beta} f(t) = \Delta\Delta_{a+\beta}^{-(1-\alpha)}\Delta_a^{-\beta} f(t) = \Delta\Delta_a^{-(1-(\alpha-\beta))} f(t) = \Delta_a^{\alpha-\beta} f(t),$$

for all $t \in \mathbb{N}_{a+\beta-\alpha}$. Now, since $\Delta_a^{-\beta} f(a+\beta-1) = 0$, we infer by Proposition 2.12 that ${}_*\Delta^{\alpha}_{a+\beta-1}\Delta_a^{-\beta} f(t) = \Delta^{\alpha}_{a+\beta-1}\Delta_a^{-\beta} f(t)$, from which the result follows.

We now compose a fractional sum with a fractional difference.

Theorem 2.26 *Assume* $f : \mathbb{N}_{a+\alpha-1} \to \mathbb{R}$ *and* $g : \mathbb{N}^{b+1-\alpha} \to \mathbb{R}$. *Moreover, let* $0 < \alpha \le 1$. *Then,*

$$\Delta_a^{-\alpha}{}_*\Delta^{\alpha}_{a+\alpha-1} f(t) = f(t) - f(a+\alpha-1), \quad t \in \mathbb{N}_{a+\alpha}.$$

Moreover,

$${}_{b-1}\Delta^{-\alpha}{}_{b+1-\alpha}\Delta^{\alpha} g(t) = g(t+1) - \frac{(b-(t+1))^{\underline{\alpha-1}}}{\Gamma(\alpha)} g(b+1-\alpha), \quad t \in \mathbb{N}^{b-\alpha-1}.$$

Proof By (2.14) and Corollary 1.15, we have

$$\Delta_a^{-\alpha}{}_*\Delta^{\alpha}_{a+\alpha-1} f(t) = \Delta_a^{-\alpha}\Delta_{a+\alpha-1}^{-(1-\alpha)}\Delta f(t) = \Delta_{a+\alpha-1}^{-1}\Delta f(t) = f(t) - f(a+\alpha-1),$$

and the first formula is proved.

Now, assume $0 < \alpha < 1$ (the case $\alpha = 1$ is immediately seen). Using Proposition 2.12, we obtain

$$\begin{aligned}
&{}_{b-1}\Delta^{-\alpha}{}_{b+1-\alpha}\Delta^{\alpha} g(t) \\
&\quad = {}_{b-1}\Delta^{-\alpha}\left[{}_{b+1-\alpha}\Delta^{\alpha}_{*} g(s) + \frac{(b-\alpha-s)^{\underline{-\alpha}}}{\Gamma(1-\alpha)} g(b+1-\alpha)\right](t).
\end{aligned}$$

From (2.15), we may conclude that

$$
\begin{aligned}
&{}_{b-1}\Delta^{-\alpha}{}_{b+1-\alpha}\Delta_*^{\alpha} g(t)\\
&\quad = -{}_{b-1}\Delta^{-\alpha}{}_{b-\alpha}\Delta^{-(1-\alpha)}\Delta g(t) = -{}_{b-\alpha}\Delta^{-1}\Delta g(t) = g(t+1) - g(b+1-\alpha).
\end{aligned}
$$

Moreover, using Theorem 2.21, we obtain

$$
\begin{aligned}
&{}_{b-1}\Delta^{-\alpha}\left[\frac{(b-\alpha-s)^{\underline{-\alpha}}}{\Gamma(1-\alpha)} g(b+1-\alpha)\right](t)\\
&= \frac{g(b+1-\alpha)}{\Gamma(1-\alpha)}{}_{b-1}\Delta^{-\alpha}\left[(b-\alpha-s)^{\underline{-\alpha}}\right](t) = \frac{g(b+1-\alpha)}{\Gamma(1-\alpha)}\\
&\quad \cdot\left({}_{b}\Delta^{-\alpha}\left[(b-\alpha-s)^{\underline{-\alpha}}\right](t) - \frac{(b-(t+1))^{\underline{\alpha-1}}\Gamma(1-\alpha)}{\Gamma(\alpha)}\right)\\
&= g(b+1-\alpha)\left[1 - \frac{(b-(t+1))^{\underline{\alpha-1}}}{\Gamma(\alpha)}\right].
\end{aligned}
$$

Finally, adding up, we obtain the desired result.

2.5 Fractional Leibniz Formula

We start by proving a simple yet useful result.

Lemma 2.27 *Suppose that* $g : \mathbb{N}_a \to \mathbb{R}$ *and* $k \in \mathbb{N}_0$, $\alpha \in \mathbb{R}$. *Then,*

$$
\sum_{n=0}^{k}(-1)^n\binom{k}{n}\Delta^n g(t-\alpha-n) = g(t-\alpha-k), \quad t \in \mathbb{N}_{a+\alpha+k}.
$$

Proof We use induction on k. So, let $k = 0$ and $t \in \mathbb{N}_{a+\alpha}$. Then, the equality trivially holds. Now, assume that the equality holds for $k \in \mathbb{N}_0$, and let $t \in \mathbb{N}_{a+\alpha+k+1}$. We have

$$
\begin{aligned}
&\sum_{n=0}^{k+1}(-1)^n\binom{k+1}{n}\Delta^n g(t-\alpha-n)\\
&\quad = g(t-\alpha) + (-1)^{k+1}\Delta^{k+1} g(t-\alpha-(k+1))\\
&\qquad + \sum_{n=1}^{k}(-1)^n\left[\binom{k}{n} + \binom{k}{n-1}\right]\Delta^n g(t-\alpha-n)
\end{aligned}
$$

$$= \sum_{n=0}^{k} (-1)^n \binom{k}{n} \Delta^n g(t-\alpha-n)$$

$$- \sum_{n=0}^{k} (-1)^n \binom{k}{n} \Delta^n \Delta g(t-1-\alpha-n) = g(t-\alpha-k) - \Delta g(t-1-\alpha-k)$$

$$= g(t-\alpha-(k+1)),$$

and the proof is done.

We are now in a position to prove the fractional Leibniz formula.

Theorem 2.28 (Fractional Leibniz Formula) *Suppose that $f, g : \mathbb{N}_a \to \mathbb{R}$ and $\alpha > 0$. Then,*

$$\Delta_a^{-\alpha}[fg](t) = \sum_{n=0}^{t-\alpha-a} \binom{-\alpha}{n} \Delta_a^{-(\alpha+n)} f(t) \Delta^n g(t-\alpha-n), \quad t \in \mathbb{N}_{a+\alpha}. \tag{2.16}$$

Proof We start by fixing $t \in \mathbb{N}_{a+\alpha}$. By Lemma 2.27, we have

$$\sum_{n=0}^{t-s-\alpha} (-1)^n \binom{t-s-\alpha}{n} \Delta^n g(t-\alpha-n) = g(s), \quad s \in \mathbb{N}_a^{t-\alpha}.$$

Therefore,

$$\Delta_a^{-\alpha}[fg](t)$$

$$= \frac{1}{\Gamma(\alpha)} \sum_{s=a}^{t-\alpha} (t-(s+1))^{\underline{\alpha-1}} f(s) g(s)$$

$$= \frac{1}{\Gamma(\alpha)} \sum_{s=a}^{t-\alpha} \sum_{n=0}^{t-s-\alpha} (t-(s+1))^{\underline{\alpha-1}} (-1)^n \binom{t-s-\alpha}{n} f(s) \Delta^n g(t-\alpha-n)$$

$$= \frac{1}{\Gamma(\alpha)} \sum_{n=0}^{t-a-\alpha} \sum_{s=a}^{t-\alpha-n} (t-(s+1))^{\underline{\alpha-1}} (-1)^n \binom{t-s-\alpha}{n} f(s) \Delta^n g(t-\alpha-n).$$

Now, using the equality in (1.7), we get from the previous deduction and after some cancellations that

$$\frac{1}{\Gamma(\alpha)} \sum_{n=0}^{t-a-\alpha} \sum_{s=a}^{t-\alpha-n} (t-(s+1))^{\underline{\alpha-1}} (-1)^n \binom{t-s-\alpha}{n} f(s) \Delta^n g(t-\alpha-n)$$

$$= \sum_{n=0}^{t-a-\alpha} \frac{(-\alpha)^{\underline{n}}}{n!} \frac{1}{\Gamma(n+\alpha)} \sum_{s=a}^{t-(\alpha+n)} (t-(s+1))^{\underline{\alpha+n-1}} f(s) \Delta^n g(t-\alpha-n)$$

$$= \sum_{n=0}^{t-a-\alpha} \binom{-\alpha}{n} \Delta_a^{-(\alpha+n)} f(t) \Delta^n g(t-\alpha-n),$$

which concludes the proof.

Example 2.29 For $0 < \alpha \leq 1$, we want to calculate $\Delta_0^\alpha[f(s)s](t)$, for $t \in \mathbb{N}_{1-\alpha}$.

Let $g(s) = s$, then $\Delta g(t) = 1$ and $\Delta^n g(t) = 0$ for all $n \in \mathbb{N}_2$. Using (2.16), we have

$$\Delta_0^{-(1-\alpha)}[f(s)s](t) = \Delta_0^{-(1-\alpha)} f(t)(t+\alpha-1) + (\alpha-1)\Delta_0^{-(2-\alpha)} f(t),$$

for all $t \in \mathbb{N}_{1-\alpha}$, and therefore, with the help of 5. in Theorem 1.3, we obtain

$$\Delta_0^\alpha[f(s)s](t) = \Delta_0^\alpha f(t)(t+\alpha) + \Delta_0^{-(1-\alpha)} f(t) + (\alpha-1)\Delta[\Delta_0^{-(2-\alpha)} f](t).$$

Using the equality in (2.14) and Theorem 1.14, we obtain $\Delta[\Delta_0^{-(2-\alpha)} f](t) = \Delta\Delta_{1-\alpha}^{-1}[\Delta_0^{-(1-\alpha)} f](t) = \Delta_0^{-(1-\alpha)} f(t)$, hence, the product formula

$$\Delta_0^\alpha[f(s)s](t) = \Delta_0^\alpha f(t)(t+\alpha) + \alpha\Delta_0^{-(1-\alpha)} f(t), \quad t \in \mathbb{N}_{1-\alpha}. \tag{2.17}$$

It is interesting to observe that the last term in (2.17) is a fractional sum and not a fractional difference. Moreover, when $\alpha = 1$, formula (2.17) becomes

$$\Delta[f(s)s](t) = \Delta f(t)(t+1) + f(t).$$

We would like to note at this point that, using Theorems 1.10 and 2.28, we may write a formula for calculating the Caputo derivative of a composition of functions. Indeed, we have the following.

Corollary 2.30 (Fractional Chain Rule) *Let f and g be as in Theorem 1.10. Then, for $0 < \alpha \leq 1$ and all $t \in \mathbb{N}_{a+1-\alpha}$, we have*

$${}_*\Delta_a^\alpha[f \circ g](t) = \sum_{n=0}^{t+\alpha-(a+1)} \binom{\alpha-1}{n} \Delta_a^{-(1-\alpha+n)} h(t) \Delta^{n+1} g(t+\alpha-(n+1)),$$

where $h : \mathbb{N}_a \to \mathbb{R}$ is defined by $h(s) = \int_0^1 f'(g(s) + r\Delta g(s))dr$.

Proof Using Theorems 1.10 and 2.28, we obtain

$$
\begin{aligned}
{}_*\Delta_a^{\alpha}[f \circ g](t) &= \Delta_a^{-(1-\alpha)}[\Delta[f \circ g]](t) \\
&= \Delta_a^{-(1-\alpha)}\left[\int_0^1 f'(g(s) + r\Delta g(s))dr\,\Delta g(s)\right](t) \\
&= \sum_{n=0}^{t-1+\alpha-a} \binom{\alpha-1}{n} \Delta_a^{-(1-\alpha+n)} h(t) \Delta^{n+1} g(t-1+\alpha-n),
\end{aligned}
$$

and the proof is done.

We finalize this section by showing how Theorem 2.28 may be used to deduce the Saalschutz formula (cf. [48, p. 49]), i.e.,

$$
\frac{(c-a)_m (c-b)_m}{(c)_m (c-a-b)_m} = {}_3F_2(a, b, -m; c, 1+a+b-c-m; 1), \tag{2.18}
$$

where the function ${}_3F_2$ above, known as a hypergeometric function, is defined by

$$
{}_3F_2(a_1, a_2, a_3; b_1, b_2; z) = \sum_{k=0}^{\infty} \frac{(a_1)_k (a_2)_k (a_3)_k}{(b_1)_k (b_2)_k} \frac{z^k}{k!},
$$

when the series on the right-hand side converges.

Consider parameters α, β, γ such that $\alpha > 0$, $\beta, \beta+\gamma \in \mathbb{R}\backslash\mathbb{N}^{-1}$. Moreover, define the function $h(t) = t^{\underline{\beta+\gamma}}$ on $t \in \mathbb{N}_{\beta+\gamma}$. By Theorem 2.17, we have

$$
\Delta_{\beta+\gamma}^{-\alpha} h(t) = \frac{\Gamma(\beta+\gamma+1)}{(t-(\beta+\gamma)-\alpha)!}(\beta+\gamma+\alpha+1)_{t-(\beta+\gamma)-\alpha}, \quad \text{for } t \in \mathbb{N}_{\alpha+\beta+\gamma}.
$$

Now consider the functions $f, g : \mathbb{N}_{\beta+\gamma} \to \mathbb{R}$ defined by

$$
f(t) = (t-\gamma)^{\underline{\beta}}, \quad g(t) = t^{\underline{\gamma}}.
$$

Then, for $t \in \mathbb{N}_{\alpha+\beta+\gamma}$ and $n \in \mathbb{N}_0^{t-(\beta+\gamma)-\alpha}$,

$$
\begin{aligned}
\Delta_{\beta+\gamma}^{-(\alpha+n)} f(t) &= \frac{\Gamma(\beta+1)}{(t-\alpha-\beta-\gamma-n)!}(\alpha+\beta+n+1)_{(t-\alpha-\beta-\gamma-n)}, \\
\Delta^n g(t) &= \gamma^{\underline{n}} t^{\underline{\gamma-n}}.
\end{aligned}
$$

Theorem 2.28 now implies that

$$\begin{aligned}\frac{\Gamma(\beta+\gamma+1)}{(t-(\beta+\gamma)-\alpha)!}&(\beta+\gamma+\alpha+1)_{t-(\beta+\gamma)-\alpha}\\ &=\sum_{n=0}^{t-\alpha-\beta-\gamma}\binom{-\alpha}{n}\frac{\Gamma(\beta+1)}{(t-\alpha-\beta-\gamma-n)!}\\ &\quad\times(\alpha+\beta+n+1)_{(t-\alpha-\beta-\gamma-n)}\gamma^{\underline{n}}(t-\alpha-n)^{\underline{\gamma-n}},\end{aligned}\tag{2.19}$$

for $t\in\mathbb{N}_{\alpha+\beta+\gamma}$.

Now, consider further that $\alpha+\beta\in\mathbb{R}\backslash\mathbb{N}^{-1}$. Put $a=\alpha$, $b=-\gamma$, $c=1+\alpha+\beta$, and $m=t-\alpha-\beta-\gamma$.

Then, (2.19) may be written as

$$\begin{aligned}\frac{\Gamma(\beta+\gamma+1)}{m!}&\frac{(\beta+\gamma+\alpha+1)_m}{\Gamma(\beta+1)}\\ &=\sum_{n=0}^{t-\alpha-\beta-\gamma}\binom{-\alpha}{n}\frac{(\alpha+\beta+n+1)_{(t-\alpha-\beta-\gamma-n)}\gamma^{\underline{n}}(t-\alpha-n)^{\underline{\gamma-n}}}{(t-\alpha-\beta-\gamma-n)!}\\ &=\frac{\Gamma(1+\alpha+\beta+m)}{\Gamma(m+\beta+1)}\sum_{n=0}^{m}\frac{(\alpha)_n}{n!}\frac{(-\gamma)_n\Gamma(m+\beta+1+\gamma-n)}{\Gamma(\alpha+\beta+n+1)(m-n)!},\end{aligned}$$

where we have used $(-1)^n(\alpha)_n=(-\alpha)^{\underline{n}}$. Now, after some rearrangements, we get

$$\begin{aligned}\frac{\Gamma(\beta+\gamma+1)}{\Gamma(1+\alpha+\beta+m)}&\frac{(\beta+\gamma+\alpha+1)_m}{\Gamma(\beta+1)}\Gamma(m+\beta+1)\\ &=\frac{\Gamma(1+\beta+\gamma+m)}{\Gamma(1+\alpha+\beta)}\sum_{n=0}^{m}\frac{(\alpha)_n(-\gamma)_n(-m)_n}{(1+\alpha+\beta)_n(-\beta-\gamma-m)_n}\frac{1}{n!},\end{aligned}$$

or

$$\begin{aligned}\frac{\Gamma(\beta+\gamma+1)\Gamma(1+\alpha+\beta)}{\Gamma(1+\alpha+\beta+m)}&\frac{(\beta+\gamma+\alpha+1)_m\Gamma(m+\beta+1)}{\Gamma(\beta+1)\Gamma(1+\beta+\gamma+m)}\\ &=\sum_{n=0}^{m}\frac{(\alpha)_n(-\gamma)_n(-m)_n}{(1+\alpha+\beta)_n(-\beta-\gamma-m)_n}\frac{1}{n!}.\end{aligned}$$

Now, note that

$$
\begin{aligned}
(c-a)_m &= \frac{\Gamma(1+\beta+m)}{\Gamma(1+\beta)} \\
(c-b)_m &= \frac{\Gamma(1+\alpha+\beta+\gamma+m)}{\Gamma(1+\alpha+\beta+\gamma)} \\
(c)_m &= \frac{\Gamma(1+\alpha+\beta+m)}{\Gamma(1+\alpha+\beta)} \\
(c-a-b)_m &= \frac{\Gamma(1+\beta+\gamma+m)}{\Gamma(1+\beta+\gamma)}.
\end{aligned}
$$

The equality (2.18) now follows from the arbitrariness of α, β, γ, and t. Note, however, that some restrictions on the parameters come with this deduction, namely, $a > 0$, $c \in \mathbb{R}\backslash\mathbb{N}^0$ with $c-a-1 \in \mathbb{R}\backslash\mathbb{N}^{-1}$, and $b \in \mathbb{R}$ with $c-a-b-1 \in \mathbb{R}\backslash\mathbb{N}^{-1}$.

2.6 A Digression into Physics: A Novel Entropic Functional

The concept of *entropy* was introduced by Clausius [17] in thermodynamics to measure the amount of energy in a system that cannot produce work, and given an atomic interpretation in the foundational works of statistical mechanics and gas dynamics by Boltzmann and Gibbs [13, 30] (see also [4]). For a probability distribution $P = \{p_1, \ldots, p_n\}$, the well-known Boltzmann–Gibbs entropy reads as

$$
S^{BG}(p) = \sum_i -\ln(p_i)p_i. \tag{2.20}
$$

Throughout the years and after the abovementioned works of Clausius and Boltzmann–Gibbs were published, many novel definitions of entropies (or generalized entropies) have appeared in the literature,[3] perhaps the most widely known one being the Tsallis entropy introduced in [49], i.e.,

$$
S_q^T(p) = \frac{1-\sum_i p_i^q}{q-1}, \quad q \neq 1. \tag{2.21}
$$

Observe that when $q \to 1$, then $S_q^T(p)$ becomes $S^{BG}(p)$. In this sense, (2.21) generalizes (2.20).

[3] We refer the reader to the review paper [4] in order to find deeper information on the subject.

From now on in this section, we consider a Liouville[4]-type Caputo fractional difference of order $0 < \alpha \le 1$:

$$
{}_*\Delta_-^{\alpha} g(t) = -\frac{1}{\Gamma(1-\alpha)} \sum_{s=t+1-\alpha}^{\infty} (s-(t+1))^{\underline{-\alpha}} \Delta g(s), \quad t \in \mathbb{N}_{\alpha-1}, \tag{2.22}
$$

where $g : \mathbb{N}_0 \to \mathbb{R}$ is a given function. With the help of this operator, we will define an entropic functional that generalizes Fermi's quadratic entropy (cf. [3]):

$$
S^F(p) = \sum_i p_i(1-p_i), \tag{2.23}
$$

where, again, $P = \{p_1, p_2, \ldots, p_n\}$ is a probability distribution. To do that, we use Abe's approach conceived in [3]. Concretely, Abe [3] noted that the Boltzmann–Gibbs entropy may be written as

$$
S^{BG}(p) = \sum_i -\frac{d}{dt} p_i^t \bigg|_{t=1}. \tag{2.24}
$$

Moreover, substituting the operator $\frac{d}{dt}$ by the Jackson q-derivative in (2.24), i.e., by $\Delta_q f(t) = \frac{f(qt)-f(t)}{(q-1)t}$, $q \neq 1$ and $t \neq 0$, we obtain the Tsallis q-entropy introduced above in (2.21). This idea was further explored in order to postulate novel (generalized) entropies, e.g., in [14, 26]. Essentially, the procedure is, in all cases, to substitute the differential operator $\frac{d}{dt}$ by another suitable one.

In this book we will use (2.22) to calculate ${}_*\Delta_-^{\alpha} g(\alpha)$ with $g(t) = p_i^t$, where we assume $0 < \alpha < 1$ and $0 < p_i < 1$, $i \in \{1, 2, \ldots, n\}$ with $n \in \mathbb{N}_2$, and then use the right-hand side of (2.24) with $-\frac{d}{dt}$ replaced by ${}_*\Delta_-^{\alpha}$ to postulate a novel entropic functional. Before we proceed, we need the following lemma.

Lemma 2.31 *For $s \in \mathbb{N}_0$ and $0 < \alpha < 1$, we have*

$$
\binom{s-\alpha}{-\alpha} = (-1)^s \binom{\alpha-1}{s}.
$$

Proof By [47, (2.3) on p. 298], the next equality holds:

$$
\Gamma(s+1-\alpha)\Gamma(\alpha-s) = (-1)^s \Gamma(1-\alpha)\Gamma(\alpha).
$$

[4] The reader is invited to check the analogy with the continuous case in [42, p. 79].

But,

$$\binom{s-\alpha}{-\alpha} = (-1)^s \binom{\alpha-1}{s} \Leftrightarrow \frac{\Gamma(s+1-\alpha)}{\Gamma(1-\alpha)} = (-1)^s \frac{\Gamma(\alpha)}{\Gamma(\alpha-s)},$$

and the result follows.

We may now proceed to calculate ${}_*\Delta_-^\alpha f(\alpha)$. We have

$$\begin{aligned}
{}_*\Delta_-^\alpha f(\alpha) &= -\frac{1}{\Gamma(1-\alpha)} \sum_{s=1}^{\infty} (s-(\alpha+1))^{\underline{-\alpha}} \Delta p_i^s \\
&= \frac{(1-p_i)p_i}{\Gamma(1-\alpha)} \sum_{s=0}^{\infty} (s-\alpha)^{\underline{-\alpha}} p_i^s \\
&= (1-p_i)p_i \sum_{s=0}^{\infty} \binom{s-\alpha}{-\alpha} p_i^s \\
&= (1-p_i)p_i \sum_{s=0}^{\infty} (-1)^s \binom{\alpha-1}{s} p_i^s \\
&= (1-p_i)^\alpha p_i,
\end{aligned}$$

where we have used the Binomial Theorem to obtain the last equality. Note that $(1-p_i)^\alpha p_i$ is also well-defined for $\alpha = 1$ and for $p_i = 0$ or $p_i = 1$ (with $0 < \alpha \leq 1$). Therefore, we postulate the following entropy:

$$S_\alpha^n(p) = \sum_{i=1}^{n} (1-p_i)^\alpha p_i, \quad 0 < \alpha \leq 1, \; p_i \in [0, 1]. \tag{2.25}$$

Remark 2.32 The entropic functional defined by (2.25) generalizes Fermi's entropy in the sense that when $\alpha = 1$, then $S_1^n = S^F$.

In what follows we will highlight some properties of the entropy defined in (2.25) (see also Fig. 2.1).

We start considering, for $0 < \alpha \leq 1$, the function $F_\alpha : [0, 1] \to \mathbb{R}_0^+$ defined by $F_\alpha(x) = (1-x)^\alpha x$. Differentiating it twice on $(0, 1)$, we get

$$F_\alpha''(x) = \alpha[(\alpha+1)x - 2](1-x)^{\alpha-2} < 0, \quad x \in (0, 1),$$

which implies that F_α is concave. It is now immediate to conclude that the entropy defined by (2.25) with $\alpha \in (0, 1]$ satisfies the following three Shannon–Khinchin axioms (cf. [4, pag. 5]), i.e., for

$$\Delta_n = \{(p_1, ..., p_n) \in \mathbb{R}^n \mid p_i \geq 0, \; p_1 + ... + p_n = 1\}, \quad n \geq 2,$$

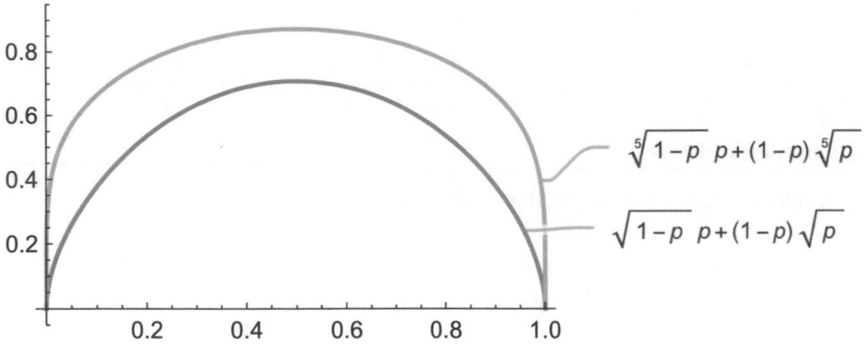

Fig. 2.1 $S^2_{1/2}(p)$ and $S^2_{1/5}(p)$

we have

1. S^n_α is nonnegative and continuous on Δ_n.
2. For all $(p_1, ..., p_n) \in \Delta_n$: $S^n_\alpha(p_1, ..., p_n) \leq S^n_\alpha(1/n, ..., 1/n)$.
3. For all $(p_1, ..., p_n) \in \Delta_n$: $S^{n+1}_\alpha(p_1, ..., p_n, 0) = S^n_\alpha(p_1, ..., p_n)$.

About 40 years ago, B. Lesche [43] introduced a way of measuring the stability of a system. Concretely, a system S over a probability distribution $P = \{p_1, \ldots, p_n\}$ with $n \geq 2$ is said to be *Lesche stable* iff

$$\forall \varepsilon > 0 \, \exists \delta > 0 \; s.t. \; \forall n, \; \forall p, p' \in P : \|p - p'\|_1 \leq \delta \Rightarrow \frac{|S(p) - S(p')|}{S_{\max}} < \varepsilon,$$

where $\|p - p'\|_1 = \sum_{i=1}^n |p_i - p'_i|$. More recently, Cao and Luo provided criteria for an entropic system to be Lesche stable [15]. Particularly,

> Let $g : [0, \infty) \to [0, \infty)$ be a concave function such that $g(0) = 0$. Put $f(x) = g(-\ln(x))$ and suppose that $xf(x)$ is concave. Then, $\sum_i p_i f(p_i)$ is Lesche stable (cf. [15, Example 3]).

We may apply the above result to show that (2.25) is Lesche stable for fixed $0 < \alpha \leq 1$. Indeed, define

$$g_\alpha(x) = \left(1 - e^{-x}\right)^\alpha, \quad x \geq 0.$$

It is clear that g is a nonnegative function with $g(0) = 0$. Moreover, differentiating g twice, it is easy to check that g is concave. Now, $f_\alpha(x) = g_\alpha(-\ln(x)) = (1-x)^\alpha$, and hence $xf_\alpha(x) = F_\alpha(x)$ is concave. Therefore, we conclude that (2.25) is Lesche stable.

The interested reader may consult [4] and the references stated therein to find out more information about generalized entropies.

2.7 Exercises

1. Complete the proof of Proposition 2.12.
2. Prove the binomial formula enunciated in Remark 2.16.
3. Evaluate the following:

   (a) $\Delta_{\frac{3}{2}}^{-1}[(s-1)^{\underline{\frac{1}{2}}}](t),\ t \in \mathbb{N}_{\frac{3}{2}}$.

   (b) $\Delta_{2}^{\frac{1}{2}}[s(s-1)](t),\ t \in \mathbb{N}_{\frac{5}{2}}$.

4. Complete the proof of Theorem 2.23
5. Compute $\Delta_0^{\alpha}[\sin(s)s^2](t), 0 < \alpha \leq 1$.
6. The discrete convolution of two functions $f, g : \mathbb{N}_a \to \mathbb{R}$, with $a \in \mathbb{R}$, denoted by $(f * g)_a$, is defined by

$$(f * g)_a(t) = \sum_{\tau=a}^{t-1} f(t-\tau-1+a)g(\tau), \quad t \in \mathbb{N}_a.$$

   (a) Show that the convolution is commutative and associative.
   (b) Compute $(e_p(t,a) * 1)_a$, where $p \neq 0, 1$.
   (c) Let $a \in \mathbb{R}$ and $\alpha \in \mathbb{R}^+ \backslash \mathbb{N}_1$. Define the functions

$$p(t) = \frac{(t-a)^{\underline{\alpha-1}}}{\Gamma(\alpha)} \text{ and } q(t) = \frac{(t-a+1-2\alpha)^{\underline{-\alpha}}}{\Gamma(1-\alpha)}, \quad t \in \mathbb{N}_{a+\alpha-1}.$$

   Show that $(p * q)_{a+\alpha-1}(t) = 1$ for all $t \in \mathbb{N}_{a+\alpha}$.

7. Show that the Tsallis entropy given by (2.21) may be obtained by calculating

$$\sum_i -\Delta_q[p_i^s](t)_{|t=1}, \quad \text{with} \sum_i p_i = 1,$$

   where $\Delta_q[f](t) = \frac{f(qt)-f(t)}{(q-1)t}$, $q \neq 1$.
8. For $0 < \alpha \leq 1$, show that the function $g_\alpha(x) = \left(1 - e^{-x}\right)^{\alpha}$, $x \geq 0$, is concave.

Chapter 3
Fractional Difference Equations

In this chapter we will be mainly concerned with solving[1] the discrete fractional initial value problem (DFIVP):

$$
{}_*\Delta^{\alpha}_{a+\alpha-1}x(t) = f(t+\alpha-1, x(t+\alpha-1)), \quad t \in \mathbb{N}_a, \tag{3.1}
$$

$$
x(a+\alpha-1) = A, \tag{3.2}
$$

for a suitable function f and real numbers a, A, α with $\alpha \in (0, 1]$.

The following result allows us to write (3.1)–(3.2) as an equivalent summation equation, which will be used throughout this chapter.

Lemma 3.1 *Consider $a, A \in \mathbb{R}$ and $0 < \alpha \leq 1$. Suppose that $f : \mathbb{N}_{a+\alpha-1} \times \mathbb{R} \to \mathbb{R}$. Then $x : \mathbb{N}_{a+\alpha-1} \to \mathbb{R}$ is a solution of (3.1)–(3.2) if and only if it is a solution of*

$$
x(t) = A + \frac{1}{\Gamma(\alpha)} \sum_{s=a}^{t-\alpha} (t-(s+1))^{\underline{\alpha-1}} f(s+\alpha-1, x(s+\alpha-1)), \ t \in \mathbb{N}_{a+\alpha-1}. \tag{3.3}
$$

Remark 3.2 Note that, for each a, A, α and a function f as in Lemma 3.1, a solution of (3.3) is unique.

Proof Suppose that the function x solves (3.1)–(3.2). Applying the operator $\Delta_a^{-\alpha}$ to both sides of (3.1) and using Theorem 2.26 we get (3.3) on $\mathbb{N}_{a+\alpha}$. The equality also holds for $t = a+\alpha-1$ because empty sums are assumed to be zero.

[1] By a solution we mean a function $x : \mathbb{N}_{a+\alpha-1} \to \mathbb{R}$ that satisfies (3.1) and (3.2).

R. A. C. Ferreira, *Discrete Fractional Calculus and Fractional Difference Equations*, SpringerBriefs in Mathematics,
https://doi.org/10.1007/978-3-030-92724-0_3

Suppose now that the function x is a solution of (3.3). It is clear that $x(a + \alpha - 1) = A$. Moreover, by Theorem 2.24, we obtain, after applying the operator ${}_*\Delta^{\alpha}_{a+\alpha-1}$ to both sides of (3.3), the equality in (3.1), which concludes the proof.

3.1 Linear Difference Equations

Fix $a, A \in \mathbb{R}$, $0 < \alpha \leq 1$, and consider $y, z : \mathbb{N}_{a+\alpha-1} \to \mathbb{R}$. We aim to solve explicitly in x the following DFIVP:

$$
\begin{aligned}
&{}_*\Delta^{\alpha}_{a+\alpha-1}x(t) = y(t+\alpha-1)x(t+\alpha-1) + z(t+\alpha-1), \; t \in \mathbb{N}_a, \\
&x(a+\alpha-1) = A.
\end{aligned}
$$

In order to accomplish it, let us introduce the following operator: for a function $y : \mathbb{N}_{a+\alpha-1} \to \mathbb{R}$, we define the operator T by

$$
\begin{aligned}
T^0_y[f](t) &= f(t), \\
T^1_y[f](t) &= T_y[f](t) = \Delta_a^{-\alpha}[y(s+\alpha-1)f(s+\alpha-1)](t), \\
T^{k+1}_y[f](t) &= T_y[T^k_y f](t), \quad k \in \mathbb{N}_1,
\end{aligned}
$$

for $t \in \mathbb{N}_{a+\alpha-1}$. It is easy to show that

$$
\forall k \in \mathbb{N}_{a+1} : \; T_y^{k-a}[f](s+\alpha-1) = 0, \; \forall s \in \mathbb{N}_a^{k-1}. \tag{3.4}
$$

Theorem 3.3 *The function*

$$
x(t) = \sum_{k=a}^{t-(\alpha-1)} \left\{ A T_y^{k-a}[1](t) + T_y^{k\ a}[\Delta_a^{-\alpha} z(s+\alpha-1)](t) \right\}, \tag{3.5}
$$

is the solution of the summation equation:

$$
x(t) = A + \Delta_a^{-\alpha}[y(s+\alpha-1)x(s+\alpha-1) + z(s+\alpha-1)](t),
$$

for all $t \in \mathbb{N}_{a+\alpha-1}$.

Proof Fix $t \in \mathbb{N}_{a+\alpha-1}$. We have that

$$
\begin{aligned}
&A + \left(\Delta_a^{-\alpha}[y(s+\alpha-1)x(s+\alpha-1) + z(s+\alpha-1)]\right)(t) \\
&= A + \sum_{s=a}^{t-\alpha} \frac{(t-(s+1))^{\underline{\alpha-1}}}{\Gamma(\alpha)} y(s+\alpha-1)x(s+\alpha-1) + \Delta_a^{-\alpha}[z(s+\alpha-1)](t)
\end{aligned}
$$

$$
\begin{aligned}
&= A + \Delta_a^{-\alpha}[z(s+\alpha-1)](t) \\
&\quad + \frac{1}{\Gamma(\alpha)} \sum_{s=a}^{t-\alpha} (t-(s+1))^{\underline{\alpha-1}} y(s+\alpha-1) \sum_{k=a}^{s} \Big[A T_y^{k-a}[1](s+\alpha-1) \\
&\qquad + T_y^{k-a}[\Delta_a^{-\alpha} z(r+\alpha-1)](s+\alpha-1) \Big] \\
&= A + \Delta_a^{-\alpha}[z(s+\alpha-1)](t) \\
&\quad + A \sum_{s=a}^{t-\alpha} \frac{1}{\Gamma(\alpha)} \sum_{k=a}^{s} (t-(s+1))^{\underline{\alpha-1}} y(s+\alpha-1) T_y^{k-a}[1](s+\alpha-1) \\
&\quad + \sum_{s=a}^{t-\alpha} \frac{1}{\Gamma(\alpha)} \sum_{k=a}^{s} (t-(s+1))^{\underline{\alpha-1}} y(s+\alpha-1) \\
&\quad \times \left(T_y^{k-a} \Delta_a^{-\alpha} z(r+\alpha-1) \right) (s+\alpha-1) \\
&= A + \Delta_a^{-\alpha}[z(s+\alpha-1)](t) \\
&\quad + A \sum_{k=a}^{t-\alpha} \frac{1}{\Gamma(\alpha)} \sum_{s=k}^{t-\alpha} (t-(s+1))^{\underline{\alpha-1}} y(s+\alpha-1) T_y^{k-a}[1](s+\alpha-1) \\
&\quad + \sum_{k-a}^{t-\alpha} \frac{1}{\Gamma(\alpha)} \sum_{s=k}^{t-\alpha} (t-(s+1))^{\underline{\alpha-1}} y(s+\alpha-1) \\
&\quad \times \left(T_y^{k-a} \Delta_a^{-\alpha} z(r+\alpha-1) \right) (s+\alpha-1) \\
&= A + A \sum_{k=a}^{t-\alpha} T_y^{k+1-a}[1](t) \\
&\quad + \Delta_a^{-\alpha}[z(s+\alpha-1)](t) + \sum_{k=a}^{t-\alpha} T_y^{k+1-a}[\Delta_a^{-\alpha} z(r+\alpha-1)](t) \\
&= \sum_{k=a}^{t-(\alpha-1)} \Big[A T_y^{k-a}[1](t) + T_y^{k-a} \Delta_a^{-\alpha}[z(s+\alpha-1)](t) \Big] = x(t),
\end{aligned}
$$

where we used Lemma 2.1 and (3.4). The uniqueness follows from Remark 3.2, and the proof is done.

Before we proceed, let us introduce some notation. We define a real-valued function E (it can be thought of as a discrete Mittag-Leffler function) by

$$E(t,r,m,a,\alpha,\beta,\lambda)=\sum_{k=a}^{t-\alpha-m}\frac{\lambda^{k-a}}{\Gamma((k-a)\alpha+\beta)}(t-r+(k-a)(\alpha-1))^{\underline{(k-a)\alpha+\beta-1}},$$

whenever the right-hand side makes sense.

Corollary 3.4 *If $y(t)=\lambda$ for some $\lambda\in\mathbb{R}$ and all $t\in\mathbb{N}_{a+\alpha-1}$ in Theorem 3.3, then the solution given by* (3.5) *is*

$$x(t)=AE(t,a+\alpha-1,-1,a,\alpha,1,\lambda)+\sum_{r=a}^{t-\alpha}E(t,r+1,r-a,a,\alpha,\alpha,\lambda)z(r+\alpha-1),$$

for all $t\in\mathbb{N}_{a+\alpha-1}$.

Proof We start by showing that

$$\sum_{k=a}^{t-(\alpha-1)}T_\lambda^{k-a}[1](t)=E(t,a+\alpha-1,-1,a,\alpha,1,\lambda),\quad t\in\mathbb{N}_{a+\alpha-1}. \tag{3.6}$$

The equality is obvious for $t=a+\alpha-1$. Therefore, we fix $t\in\mathbb{N}_{a+\alpha}$, and we will use induction on k to show that

$$T_\lambda^{k-a}[1](t)=\frac{\lambda^{k-a}}{\Gamma((k-a)\alpha+1)}(t-(a+\alpha-1)+(k-a)(\alpha-1))^{\underline{(k-a)\alpha}}, \tag{3.7}$$

for all $k\in\mathbb{N}_a^{t-(\alpha-1)}$. The case $k=a$ is obvious. Assume that (3.7) holds for $k\in\mathbb{N}_a^{t-1-(\alpha-1)}$. Then,

$$\begin{aligned}
T_\lambda^{k+1-a}[1](t)&=T_\lambda[T_\lambda^{k-a}[1](s+\alpha-1)](t)\\
&=\frac{\lambda^{k+1-a}}{\Gamma((k-a)\alpha+1)}\Delta_k^{-\alpha}[(s+\alpha-1-(a+\alpha-1)+(k-a)(\alpha-1))^{\underline{(k-a)\alpha}}](t)\\
&=\frac{\lambda^{k+1-a}}{\Gamma((k-a)\alpha+1)}\sum_{s=k}^{t-\alpha}(t-(s+1))^{\underline{\alpha-1}}(s-a+(k-a)(\alpha-1))^{\underline{(k-a)\alpha}}\\
&=\frac{\lambda^{k+1-a}}{\Gamma((k-a)\alpha+1)}\sum_{s=k}^{t-\alpha}(t-(s+1))^{\underline{\alpha-1}}(s-k+(k-a)\alpha)^{\underline{(k-a)\alpha}}
\end{aligned}$$

$$= \frac{\lambda^{k+1-a}}{\Gamma((k-a)\alpha+1)} \sum_{s=k+(k-a)\alpha}^{t+(k-a)\alpha-\alpha} (t+(k-a)\alpha-(s+1))^{\underline{\alpha-1}}(s-k)^{\underline{(k-a)\alpha}}$$

$$= \frac{\lambda^{k+1-a}}{\Gamma((k+1-a)\alpha+1)}(t-a+(k-a)(\alpha-1))^{\underline{(k+1-a)\alpha}},$$

where we have used (3.4). Equality (3.7) is therefore proved.

To complete the proof, we will also show that

$$\sum_{k=a}^{t-(\alpha-1)} T_\lambda^{k-a}[\Delta_a^{-\alpha} z(s+\alpha-1)](t) = \sum_{r=a}^{t-\alpha} E(t, r+1, r-a, a, \alpha, \alpha, \lambda) z(r+\alpha-1).$$

Similarly as above, we start by fixing $t \in \mathbb{N}_{a+\alpha}$, and we use induction on $k \in \mathbb{N}_a^{t-(\alpha-1)}$ to prove that

$$T_\lambda^{k-a}[\Delta_a^{-\alpha} z(s+\alpha-1)](t)$$
$$= \sum_{r=a}^{t-\alpha-(k-a)} \frac{\lambda^{k-a}}{\Gamma((k-a)\alpha+\alpha)}$$
$$\times (t-(r+1)+(k-a)(\alpha-1))^{\underline{(k-a)\alpha+\alpha-1}} z(r+\alpha-1).$$

For $k = a$, the equality holds since

$$\Delta_a^{-\alpha} z(s+\alpha-1)(t) = \sum_{r=a}^{t-\alpha} \frac{(t-(r+1))^{\underline{\alpha-1}}}{\Gamma(\alpha)} z(r+\alpha-1).$$

Now, suppose that it holds for some $k \in \mathbb{N}_a^{t-1-(\alpha-1)}$. Then,

$$T_\lambda^{k+1-a}[\Delta_a^{-\alpha} z(s+\alpha-1)](t) = \Delta_k^{-\alpha} \lambda T_\lambda^{k-a} \Delta_a^{-\alpha} z(s+\alpha-1)(t)$$
$$= \frac{1}{\Gamma(\alpha)} \sum_{n=k}^{t-\alpha} (t-(n+1))^{\underline{\alpha-1}}$$
$$\cdot \sum_{r=a}^{n-1-(k-a)} \frac{\lambda^{k+1-a}}{\Gamma((k-a)\alpha+\alpha)}$$
$$\times (n+\alpha-1-(r+1)+(k-a)(\alpha-1))^{\underline{(k-a)\alpha+\alpha-1}} z(r+\alpha-1)$$
$$= \sum_{n=a}^{t-(k-a)-\alpha} (t-(n+k-a+1))^{\underline{\alpha-1}} \frac{\lambda^{k+1-a}}{\Gamma((k-a)\alpha+\alpha)}$$

$$
\begin{aligned}
&\cdot \frac{1}{\Gamma(\alpha)} \sum_{r=a}^{n-1} (n+k-a+\alpha-1-(r+1)+(k-a)(\alpha-1))^{\underline{(k-a)\alpha+\alpha-1}} \\
&\times z(r+\alpha-1) \\
= &\sum_{r=a}^{t-(k-a)-\alpha} \frac{\lambda^{k+1-a}}{\Gamma((k-a)\alpha+\alpha)} z(r+\alpha-1) \\
&\cdot \frac{1}{\Gamma(\alpha)} \sum_{n=r+1}^{t-(k-a)-\alpha} (t-(n+k-a+1))^{\underline{\alpha-1}} \\
&\times (n+k-a+\alpha-1-(r+1)+(k-a)(\alpha-1))^{\underline{(k-a)\alpha+\alpha-1}} \\
= &\sum_{r=a}^{t-(k+1-a)-\alpha} \frac{\lambda^{k+1-a}}{\Gamma((k-a)\alpha+\alpha)} z(r+\alpha-1) \\
&\cdot \frac{1}{\Gamma(\alpha)} \sum_{n=r+1+(k-a)\alpha+\alpha-1}^{t+(k+1-a)(\alpha-1)-\alpha} (t+(k+1-a)(\alpha-1)-(n+1))^{\underline{\alpha-1}} \\
&\times (n-(r+1))^{\underline{(k-a)\alpha+\alpha-1}} \\
= &\sum_{r=a}^{t-(k+1-a)-\alpha} \frac{\lambda^{k+1-a}}{\Gamma((k+1-a)\alpha+\alpha)} \\
&\times (t-(r+1)+(k+1-a)(\alpha-1))^{\underline{(k+1-a)\alpha+\alpha-1}} z(r+\alpha-1).
\end{aligned}
$$

We have, therefore, obtained

$$
\begin{aligned}
&\sum_{k=a}^{t-(\alpha-1)} T_{\lambda}^{k-a}[\Delta_a^{-\alpha} z(s+\alpha-1)](t) \\
&= \sum_{k=a}^{t-(\alpha-1)} \sum_{r=a}^{t-\alpha-(k-a)} \frac{\lambda^{k-a}}{\Gamma((k-a)\alpha+\alpha)} (t-(r+1)+(k-a)(\alpha-1))^{\underline{(k-a)\alpha+\alpha-1}} \\
&\times z(r+\alpha-1),
\end{aligned}
$$

which is equal to

$$
\begin{aligned}
&\sum_{k=a}^{t-\alpha} \sum_{r=a}^{t-\alpha-(k-a)} \frac{\lambda^{k-a}}{\Gamma((k-a)\alpha+\alpha)} (t-(r+1)+(k-a)(\alpha-1))^{\underline{(k-a)\alpha+\alpha-1}} \\
&\times z(r+\alpha-1)
\end{aligned}
$$

$$
= \sum_{r=a}^{t-\alpha} \underbrace{\sum_{k=a}^{t-\alpha-(r-a)} \frac{\lambda^{k-a}}{\Gamma((k-a)\alpha+\alpha)} (t-(r+1)+(k-a)(\alpha-1))^{\underline{(k-a)\alpha+\alpha-1}}}_{E(t,r+1,r-a,a,\alpha,\alpha,\lambda)}
$$

$$
\times z(r+\alpha-1).
$$

The proof is done.

It is pertinent to formulate the following consequence of Corollary 3.4.

Corollary 3.5 *Suppose that the function z is a constant equal to K in Corollary 3.4. Then,*

$$
x(t) = AE(t, a+\alpha-1, -1, a, \alpha, 1, \lambda) + KE(t, a, 0, a, \alpha, \alpha+1, \lambda), \quad t \in \mathbb{N}_{a+\alpha-1}.
$$

Proof We just need to calculate $\sum_{r=a}^{t-\alpha} E(t, r+1, r-a, a, \alpha, \alpha, \lambda)$. We have

$$
\begin{aligned}
&\sum_{r=a}^{t-\alpha} E(t, r+1, r-a, a, \alpha, \alpha, \lambda) \\
&= \sum_{r=a}^{t-\alpha} \sum_{k=a}^{t-\alpha-(r-a)} \frac{\lambda^{k-a}}{\Gamma((k-a)\alpha+\alpha)} (t-(r+1)+(k-a)(\alpha-1))^{\underline{(k-a)\alpha+\alpha-1}} \\
&= \sum_{k=a}^{t-\alpha} \frac{\lambda^{k-a}}{\Gamma((k-a)\alpha+\alpha)} \sum_{r=a}^{t-\alpha-(k-a)} (t-(r+1)+(k-a)(\alpha-1))^{\underline{(k-a)\alpha+\alpha-1}} \\
&= \sum_{k=a}^{t-\alpha} \frac{\lambda^{k-a}}{\Gamma((k-a)\alpha+\alpha+1)} (t-a+(k-a)(\alpha-1))^{\underline{(k-a)\alpha+\alpha}} \\
&= E(t, a, 0, a, \alpha, \alpha+1, \lambda).
\end{aligned}
$$

The proof is done.

It follows an important special case of Corollary 3.5.

Theorem 3.6 *Let $\lambda \in \mathbb{R}$. Then the solution of the following initial value problem:*

$$
\begin{aligned}
&{}_*\Delta^{\alpha}_{a+\alpha-1} x(t) = \lambda x(t+\alpha-1), \ t \in \mathbb{N}_a, \\
&x(a+\alpha-1) = 1,
\end{aligned}
$$

is given by

$$x(t) = E(t, a+\alpha-1, -1, a, \alpha, 1, \lambda)$$
$$= \sum_{k=a}^{t-\alpha+1} \frac{\lambda^{k-a}}{\Gamma((k-a)\alpha+1)} (t-(a+\alpha-1)+(k-a)(\alpha-1))^{\underline{(k-a)\alpha}}.$$

Remark 3.7 It follows from equality (1.6) that if we let $\alpha = 1$ in Theorem 3.6, we obtain

$$E(t, a, -1, a, 1, 1, \lambda) = (1+\lambda)^{t-a}.$$

In this sense the discrete Mittag-Leffler function is a generalization of the discrete exponential function (cf. Example 1.26).

3.1.1 Two Inequalities

In this section we prove discrete fractional versions of two classical inequalities: The Gronwall's inequality and the Bernoulli's inequality.

We start by proving a comparison result for the fractional summation operator.

Lemma 3.8 *Let $a \in \mathbb{R}$, $0 < \alpha \le 1$, and $f(t, x) : \mathbb{N}_{a+\alpha-1} \times \mathbb{R} \to \mathbb{R}$ be nondecreasing in x for each $t \in \mathbb{N}_{a+\alpha-1}$. If $v, w : \mathbb{N}_{a+\alpha-1} \to \mathbb{R}$ are functions satisfying*

$$w(t) \ge A + \Delta_a^{-\alpha}[f(s+\alpha-1, w(s+\alpha-1))](t), \quad t \in \mathbb{N}_{a+\alpha-1},$$

$$v(t) \le B + \Delta_a^{-\alpha}[f(s+\alpha-1, v(s+\alpha-1))](t), \quad t \in \mathbb{N}_{a+\alpha-1},$$

where $A \ge B$ are two real numbers, then $w(t) \ge v(t)$ for all $t \in \mathbb{N}_{a+\alpha-1}$.

Proof Suppose that there exists $t_0 \in \mathbb{N}_{a+\alpha-1}$ such that $w(t_0) < v(t_0)$. Define the number $t_1 = \min\{t \in \mathbb{N}_{a+\alpha-1} : w(t) < v(t)\}$. Then,

$$w(t_1) \ge A + \frac{1}{\Gamma(\alpha)} \sum_{s=a}^{t_1-\alpha} (t_1-(s+1))^{\underline{\alpha-1}} f(s+\alpha-1, w(s+\alpha-1)),$$

$$v(t_1) \le B + \frac{1}{\Gamma(\alpha)} \sum_{s=a}^{t_1-\alpha} (t_1-(s+1))^{\underline{\alpha-1}} f(s+\alpha-1, v(s+\alpha-1)),$$

from which it follows that

$$
\begin{aligned}
w(t_1) - v(t_1) &\geq A - B \\
&\quad + \frac{1}{\Gamma(\alpha)} \sum_{s=a}^{t_1-\alpha} (t_1 - (s+1))^{\underline{\alpha-1}} \\
&\quad \times [f(s+\alpha-1, w(s+\alpha-1)) - f(s+\alpha-1, v(s+\alpha-1))] \\
&\geq 0.
\end{aligned}
$$

This is a contradiction and, therefore, the lemma is proved.

One of the most important inequalities in the theory of differential equations is known as the *Gronwall's inequality*. It was published in 1919 in the work by Gronwall [37]. A discrete version of the Gronwall's inequality seems to have appeared first in the work of Mikeladze [45]. In the next result we prove a (discrete) fractional version of it.

Theorem 3.9 (Fractional Gronwall's Inequality) *Let $a, b \in \mathbb{R}$ and $0 < \alpha \leq 1$. Suppose that $u, y : \mathbb{N}_{a+\alpha-1} \to \mathbb{R}$ are two functions with y nonnegative. If*

$$
u(t) \leq b + \Delta_a^{-\alpha}[y(s+\alpha-1)u(s+\alpha-1)](t), \quad t \in \mathbb{N}_{a+\alpha-1},
$$

then

$$
u(t) \leq b \sum_{k=a}^{t-(\alpha-1)} T_y^{k-a}[1](t), \quad t \in \mathbb{N}_{a+\alpha-1}. \tag{3.8}
$$

Proof Let x be the solution of the equation:

$$
x(t) = b + \Delta_a^{-\alpha}[y(s+\alpha-1)x(s+\alpha-1)](t),
$$

which, by Theorem 3.3, is

$$
x(t) = b \sum_{k=a}^{t-(\alpha-1)} T_y^{k-a}[1](t).
$$

Since y is nonnegative, then $f(t, x) = y(t)x$ is nondecreasing in x. Now an application of Lemma 3.8 immediately yields $u(t) \leq x(t)$, and the proof is done.

Remark 3.10 The fractional Gronwall's inequality (3.8) with $y(t) = \lambda \in \mathbb{R}$ reads as (recall (3.6))

$$
u(t) \leq bE(t, a+\alpha-1, -1, a, \alpha, 1, \lambda), \quad t \in \mathbb{N}_{a+\alpha-1}. \tag{3.9}
$$

Corollary 3.11 (Gronwall's Inequality) *Suppose that* $\alpha = 1$ *in Theorem 3.9. Then,*

$$u(t) \leq b + \sum_{s=a}^{t-1} y(s)u(s), \quad t \in \mathbb{N}_a,$$

implies

$$u(t) \leq b \prod_{s=a}^{t-1} (1 + y(s)), \quad t \in \mathbb{N}_a.$$

Proof We only have to show the validity of the following equality:

$$\sum_{k=a}^{t} T_y^{k-a}[1](t) = \prod_{s=a}^{t-1} (1 + y(s)), \quad t \in \mathbb{N}_a. \tag{3.10}$$

To do that, we start with two important observations:

$$T_y^{t+1-a}[1](t) = 0, \quad t \in \mathbb{N}_a,$$
$$T_y^{k-a}[1](t+1) = T_y^{k-a}[1](t) + y(t)T_y^{k-1-a}[1](t), \quad k \in \mathbb{N}_{a+1}^{t+1}.$$

We have

$$\begin{aligned}\Delta \sum_{k=a}^{t} T_y^{k-a}[1](t) &= \sum_{k=a+1}^{t+1} T_y^{k-a}[1](t+1) - \sum_{k=a+1}^{t} T_y^{k-a}[1](t) \\ &= y(t) \sum_{k=a+1}^{t} T_y^{k-1-a}[1](t) + y(t)T_y^{t-a}[1](t) \\ &= y(t) \sum_{k=a}^{t} T_y^{k-a}[1](t).\end{aligned}$$

By Theorem 1.28 (note that $\sum_{k=a}^{a} T_y^{k-a}[1](a) = \prod_{s=a}^{a-1}(1+y(s)) = 1$), the equality (3.10) follows and the proof is done.

We turn now into the Bernoulli's inequality. We start with the following comparison result.

Lemma 3.12 *Assume* $0 < \alpha < 1$. *Let* $x, w : \mathbb{N}_{a+\alpha-1} \to \mathbb{R}$ *be functions,* $w(t) \geq -\alpha$, *and* x *be a solution of*

$$\Delta_{a+\alpha-1}^{\alpha} x(t) = w(t+\alpha-1)x(t+\alpha-1), \quad t \in \mathbb{N}_a \tag{3.11}$$

satisfying $x(a+\alpha-1) \geq 0$. *Then,* $x(t) \geq 0$ *for all* $t \in \mathbb{N}_{a+\alpha-1}$.

Proof Recall Remark 2.9 to write

$$\Delta^{\alpha}_{a+\alpha-1}x(t) = \frac{1}{\Gamma(-\alpha)} \sum_{k=a+\alpha-1}^{t+\alpha} (t-(k+1))^{\underline{-\alpha-1}} x(k)$$

$$= x(t+\alpha) - \alpha x(t+\alpha-1) - \frac{\alpha(-\alpha+1)}{2} x(t+\alpha-2)$$

$$- \ldots - \frac{\alpha(-\alpha+1)\ldots(-\alpha+t-a)}{(t-a+1)!} x(a+\alpha-1).$$

From (3.11), we get

$$x(t+\alpha) = [w(t+\alpha-1)+\alpha]x(t+\alpha-1) + \frac{\alpha(-\alpha+1)}{2} x(t+\alpha-2)$$

$$+ \ldots + \frac{\alpha(-\alpha+1)\ldots(-\alpha+t-a)}{(t-a+1)!} x(a+\alpha-1), \; t \in \mathbb{N}_a.$$

Now, by the hypotheses, it follows that $x(t+\alpha) \geq 0$ for all $t \in \mathbb{N}_a$, which concludes the proof.

Remark 3.13 The previous lemma clearly holds for $\alpha = 1$ (just recall the definition of the discrete exponential function).

Analogously to the results deduced in Sect. 3.1, one may show (cf. [21]) that $E(t, a, -1, a, \alpha, \alpha, c)$ is the solution of

$$\Delta^{\alpha}_{a+\alpha-1}x(t) = cx(t+\alpha-1), \quad x(a+\alpha-1) = 1.$$

By Lemma 3.12 and Remark 3.13, for a real number a, $0 < \alpha \leq 1$, and $c \geq -\alpha$:

$$E(t, a, -1, a, \alpha, \alpha, c) \geq 0, \quad t \in \mathbb{N}_{a+\alpha-1}. \tag{3.12}$$

It follows the (discrete) fractional version of the Bernoulli's inequality.

Theorem 3.14 (Generalized Bernoulli's Inequality) *Let* $\alpha \in (0, 1]$, $c \in [-\alpha, \infty)$, *and* $a \in \mathbb{R}$. *Then, the following inequality holds:*

$$cE(t, a, 0, a, \alpha, \alpha+1, c) \geq c\frac{(t-a)^{\underline{\alpha}}}{\Gamma(\alpha+1)}, \quad t \in \mathbb{N}_{a+\alpha-1}. \tag{3.13}$$

Proof Let $c \geq -\alpha$ and $x : \mathbb{N}_{a+\alpha-1} \to \mathbb{R}$ be the function defined by

$$x(t) = c\frac{(t-a)^{\underline{\alpha}}}{\Gamma(\alpha+1)}.$$

Then $x(a+\alpha-1)=0$ and ${}_*\Delta^{\alpha}_{a+\alpha-1}x(t)=c$ by (2.13). Therefore,

$$cx(t+\alpha-1)+c=c^2\frac{(t+\alpha-1-a)^{\underline{\alpha}}}{\Gamma(\alpha+1)}+c\geq {}_*\Delta^{\alpha}_{a+\alpha-1}x(t).$$

Define the function m by

$$m(t+\alpha-1)=cx(t+\alpha-1)+c-{}_*\Delta^{\alpha}_{a+\alpha-1}x(t),$$

which is nonnegative. By Corollary 3.4, we get (note that $x(a+\alpha-1)=0$)

$$x(t)=\sum_{r=a}^{t-\alpha}E(t,r+1,r-a,a,\alpha,\alpha,c)(c-m(r+\alpha-1)).$$

Hence,

$$\begin{aligned}
x(t)&=cE(t,a,0,a,\alpha,\alpha+1,c)-\sum_{r=a}^{t-\alpha}m(r+\alpha-1)\\
&\quad\cdot\sum_{k=a}^{t-\alpha-(r-a)}\frac{c^{k-a}}{\Gamma((k-a)\alpha+\alpha)}(t-(r+1)+(k-a)(\alpha-1))^{\underline{(k-a)\alpha+\alpha-1}}\\
&=cE(t,a,0,a,\alpha,\alpha+1,c)-\sum_{r=0}^{t-a-\alpha}m(r+a+\alpha-1)\\
&\quad\cdot\sum_{k=a}^{t-(r+1)-(\alpha-1)}\frac{c^{k-a}}{\Gamma((k-a)\alpha+\alpha)}\\
&\quad\times(t-(r+1)-a+(k-a)(\alpha-1))^{\underline{(k-a)\alpha+\alpha-1}}\\
&=cE(t,a,0,a,\alpha,\alpha+1,c)\\
&\quad-\sum_{r=0}^{t-a-\alpha}m(r+a+\alpha-1)E(t-(r+1),a,-1,a,\alpha,\alpha,c).
\end{aligned}$$

Finally, by (3.12) we conclude that

$$x(t)\leq cE(t,a,0,a,\alpha,\alpha+1,c),$$

which is equivalent to

$$c\frac{(t-a)^{\underline{\alpha}}}{\Gamma(\alpha+1)}\leq cE(t,a,0,a,\alpha,\alpha+1,c).$$

The proof is done.

We finish this section by showing that inequality (3.13) generalizes the Bernoulli's inequality. Indeed, let $a = 0$ and $\alpha = 1$. Then, for $c \geq -1$ we have

$$\begin{aligned} cE(t, 0, 0, 0, 1, 2, c) &\geq c\frac{t^{\underline{1}}}{\Gamma(2)} \\ \Leftrightarrow \sum_{k=0}^{t-1} \frac{c^{k+1}}{\Gamma(k+2)} t^{\underline{k+1}} &\geq ct \\ \Leftrightarrow \sum_{k=0}^{t} \frac{c^{k}}{\Gamma(k+1)} t^{\underline{k}} &\geq 1 + ct \\ \Leftrightarrow (1+c)^t &\geq 1 + ct, \quad t \in \mathbb{N}_0, \end{aligned}$$

and the last inequality is the well-known Bernoulli's inequality.

3.1.2 Asymptotic Behavior

In this section we consider the following linear fractional difference equation:

$$_*\Delta^{\alpha}_{a+\alpha-1} x(t) = c(t+\alpha-1)x(t+\alpha-1), \quad t \in \mathbb{N}_a,\ 0 < \alpha \leq 1, \tag{3.14}$$

for a certain function $c : \mathbb{N}_{a+\alpha-1} \to \mathbb{R}$. We wish to study the behavior of the solutions of (3.14) as $t \to \infty$.

The following result will prove useful in what follows.

Lemma 3.15 *Assume that* $0 < \alpha \leq 1$ *and* $a \in \mathbb{R}$. *Then,*

$$\lim_{t\to\infty} (t-a)^{\underline{\alpha}} = \infty,$$

where t *is understood to belong to* $\mathbb{N}_{a+\alpha}$.

Proof We have

$$(t-a)^{\underline{\alpha}} = \frac{\Gamma(t-a+1)}{\Gamma(t-a+1-\alpha)} = (t-a)\frac{\Gamma(t-a)}{\Gamma(t-a+1-\alpha)} \geq (t-a)\frac{1}{(t-a)^{1-\alpha}},$$

where the inequality above follows from the Wendel's inequality (cf. [50, inequality (7)]). Hence, $\lim_{t\to\infty}(t-a)^{\underline{\alpha}} \geq \lim_{t\to\infty}(t-a)^{\alpha} = \infty$, and the lemma is proved.

The following comparison theorem plays an important role in proving the main results of this section.

Theorem 3.16 *Let* $0 < \alpha < 1$. *Assume that* $c_1(t) \geq c_2(t) \geq -\alpha$ *for* $t \in \mathbb{N}_{a+\alpha-1}$ *and* x, y *are solutions of*

$$ {}_{*}\Delta^{\alpha}_{a+\alpha-1}x(t) = c_1(t+\alpha-1)x(t+\alpha-1), \quad t \in \mathbb{N}_a, \tag{3.15}$$

and

$$ {}_{*}\Delta^{\alpha}_{a+\alpha-1}y(t) = c_2(t+\alpha-1)y(t+\alpha-1), \quad t \in \mathbb{N}_a, \tag{3.16}$$

respectively, with $x(a+\alpha-1) \geq y(a+\alpha-1) \geq 0$. *Then,*

$$x(t) \geq y(t) \geq 0, \quad t \in \mathbb{N}_{a+\alpha-1}.$$

Proof Using Remark 2.9 and Proposition 2.12, we may write

$$\begin{aligned}
&{}_{*}\Delta^{\alpha}_{a+\alpha-1}x(t) \\
&= \frac{1}{\Gamma(-\alpha)} \sum_{s=a+\alpha-1}^{t+\alpha} (t-(s+1))^{\underline{-\alpha-1}}x(s) - \frac{(t-(a+\alpha-1))^{\underline{-\alpha}}}{\Gamma(1-\alpha)}x(a+\alpha-1) \\
&= x(t+\alpha) - \alpha x(t+\alpha-1) - \frac{\alpha(-\alpha+1)}{2}x(t+\alpha-2) - \ldots \\
&\quad - \frac{\alpha(-\alpha+1)\ldots(-\alpha+t-a-1)}{(t-a)!}x(a+\alpha) \\
&\quad - \frac{\Gamma(-\alpha+t-a+1)}{\Gamma(1-\alpha)\Gamma(t-a+1)}x(a+\alpha-1),
\end{aligned}$$

where we have used

$$\frac{(t-(a+\alpha))^{\underline{-\alpha-1}}}{\Gamma(-\alpha)} - \frac{(t-(a+\alpha-1))^{\underline{-\alpha}}}{\Gamma(1-\alpha)} = -\frac{(t-(a+\alpha))^{\underline{-\alpha}}}{\Gamma(1-\alpha)}.$$

Using (3.15), we obtain

$$\begin{aligned}
x(t+\alpha) = [c_1(t+\alpha-1+\alpha)]x(t+\alpha-1) &+ \frac{\alpha(-\alpha+1)}{2}x(t+\alpha-2) + \ldots \\
&+ \frac{\alpha(-\alpha+1)\ldots(-\alpha+t-a-1)}{(t-a)!}x(a+\alpha) \\
&+ \frac{\Gamma(-\alpha+t-a+1)}{\Gamma(1-\alpha)\Gamma(t-a+1)}x(a+\alpha-1).
\end{aligned}$$

In an analogous way, by using (3.16), we obtain

$$
\begin{aligned}
y(t+\alpha) &= [c_2(t+\alpha-1+\alpha)]y(t+\alpha-1) + \frac{\alpha(-\alpha+1)}{2}y(t+\alpha-2) + \dots \\
&\quad + \frac{\alpha(-\alpha+1)\dots(-\alpha+t-a-1)}{(t-a)!}y(a+\alpha) \\
&\quad + \frac{\Gamma(-\alpha+t-a+1)}{\Gamma(1-\alpha)\Gamma(t-a+1)}y(a+\alpha-1).
\end{aligned}
$$

Note that $x(t)$ and $y(t)$ are nonnegative for $t \in \mathbb{N}_{a+\alpha}$. Moreover,

$$
\begin{aligned}
&x(t+\alpha) - y(t+\alpha) \\
&\quad = [c_1(t+\alpha-1+\alpha)]x(t+\alpha-1) - [c_2(t+\alpha-1+\alpha)]y(t+\alpha-1) \\
&\quad\quad + \frac{\alpha(-\alpha+1)}{2}[x(t+\alpha-2) - y(t+\alpha-2)] + \dots \\
&\quad\quad + \frac{\alpha(-\alpha+1)\dots(-\alpha+t-a-1)}{(t-a)!}[x(a+\alpha) - y(a+\alpha)] \\
&\quad\quad + \frac{\Gamma(-\alpha+t-a+1)}{\Gamma(1-\alpha)\Gamma(t-a+1)}[x(a+\alpha-1) - y(a+\alpha-1)], \quad t \in \mathbb{N}_a.
\end{aligned}
$$

The desired inequality now follows by the hypothesis.

The following result is the *Caputo version* of Lemma 3.12 and it is an immediate consequence of Theorem 3.16.

Corollary 3.17 *Assume that* $0 < \alpha < 1$ *and* $c(t) \geq -\alpha$ *for all* $t \in \mathbb{N}_{a+\alpha-1}$. *Then the solution of*

$$
{}_*\Delta^{\alpha}_{a+\alpha-1}x(t) = c(t+\alpha-1)x(t+\alpha-1), \quad t \in \mathbb{N}_a,
$$

with $x(a+\alpha-1) \geq 0$, *is nonnegative.*

Remark 3.18 We observe that, by Theorem 3.6, if β is a number such that $\beta \geq -\alpha$ for $0 < \alpha < 1$, then

$$
\sum_{k=a}^{t-\alpha+1} \frac{\beta^{k-a}}{\Gamma((k-a)\alpha+1)}(t-(a+\alpha-1)+(k-a)(\alpha-1))^{\underline{(k-a)\alpha}} \geq 0,
$$

for all $t \in \mathbb{N}_{a+\alpha-1}$. Note that the above inequality is by no means obvious when $\beta < 0$.

Theorem 3.19 *Assume that* $0 < \alpha < 1$ *and* $c(t) \geq \lambda \geq -\alpha$ *for all* $t \in \mathbb{N}_{a+\alpha-1}$. *Then the solution of*

$$_*\Delta^{\alpha}_{a+\alpha-1}x(t) = c(t+\alpha-1)x(t+\alpha-1), \quad t \in \mathbb{N}_a$$

with $x(a+\alpha-1) \geq 0$ *satisfies*

$$x(t) \geq \frac{x(a+\alpha-1)}{2}E(t, a+\alpha-1, -1, a, \alpha, 1, \lambda), \; t \in \mathbb{N}_{a+\alpha-1}.$$

Proof We first recall from Theorem 3.6 that $E(t, a+\alpha-1, -1, a, \alpha, 1, \lambda)$ solves $_*\Delta^{\alpha}_{a+\alpha-1}z(t) = \lambda z(t+\alpha-1)$ with $z(a+\alpha-1) = 1$. Now, we define

$$y(t) = \frac{x(a+\alpha-1)}{2}E(t, a+\alpha-1, -1, a, \alpha, 1, \lambda), \quad t \in \mathbb{N}_{a+\alpha-1}.$$

Then, x and y satisfy

$$_*\Delta^{\alpha}_{a+\alpha-1}x(t) = c(t+\alpha-1)x(t+\alpha-1), \quad t \in \mathbb{N}_a$$

and

$$_*\Delta^{\alpha}_{a+\alpha-1}y(t) = \lambda y(t+\alpha-1), \quad t \in \mathbb{N}_a,$$

respectively. Moreover,

$$x(a+\alpha-1) > \frac{x(a+\alpha-1)}{2}E(a+\alpha-1, a+\alpha-1, -1, a, \alpha, 1, \lambda) = y(a+\alpha-1).$$

The result now follows from Theorem 3.16.

Theorem 3.20 *Suppose that* $\lambda > 0$. *Then,*

$$\lim_{t\to\infty} E(t, a+\alpha-1, -1, a, \alpha, 1, \lambda) = \infty.$$

Proof By Theorem 3.6, we have

$$\begin{aligned} &E(t, a+\alpha-1, -1, a, \alpha, 1, \lambda) \\ &= \sum_{k=a}^{t-\alpha+1} \frac{\lambda^{k-a}}{\Gamma((k-a)\alpha+1)}(t-(a+\alpha-1)+(k-a)(\alpha-1))^{\underline{(k-a)\alpha}} \\ &\geq \frac{\lambda}{\Gamma(\alpha+1)}(t-a)^{\underline{\alpha}}, \end{aligned}$$

for all $t \in \mathbb{N}_{a+\alpha}$. The conclusion follows from Lemma 3.15.

From Theorems 3.19 and 3.20, we obtain the following result.

Theorem 3.21 *Assume that* $0 < \alpha < 1$ *and* $c(t) \geq \lambda > 0$ *for all* $t \in \mathbb{N}_{a+\alpha-1}$. *Then the solution of*

$$_*\Delta^{\alpha}_{a+\alpha-1}x(t) = c(t+\alpha-1)x(t+\alpha-1), \quad t \in \mathbb{N}_a$$

with $x(a+\alpha-1) > 0$ *satisfies* $\lim_{t\to\infty} x(t) = \infty$.

We now proceed to address the analysis in which $\lambda < 0$ in the previous theorem. We start with the following preliminary result.

Theorem 3.22 *Suppose that* $-\alpha \leq \lambda < 0$. *Then,*

$$\lim_{t\to\infty} E(t, a+\alpha-1, -1, a, \alpha, 1, \lambda) = 0.$$

Proof Let us start by recalling that (cf. (3.12))

$$E(t, a, -1, a, \alpha, \alpha, \lambda) \geq 0, \quad t \in \mathbb{N}_{a+\alpha-1}.$$

Hence,

$$\begin{aligned}
&\Delta E(t, a+\alpha-1, -1, a, \alpha, 1, \lambda) \\
&= \Delta \sum_{k=a}^{t-\alpha+1} \frac{\lambda^{k-a}}{\Gamma((k-a)\alpha+1)}(t-(a+\alpha-1)+(k-a)(\alpha-1))^{\underline{(k-a)\alpha}} \\
&= \sum_{k=a}^{t-\alpha+1} \frac{\lambda^{k-a}}{\Gamma((k-a)\alpha+1)}\Delta(t-(a+\alpha-1) \\
&\quad + (k-a)(\alpha-1))^{\underline{(k-a)\alpha}} + \lambda^{t-\alpha+2-a} \\
&= \sum_{k=a+1}^{t-\alpha+2} \frac{\lambda^{k-a}}{\Gamma((k-a)\alpha)}(t-(a+\alpha-1)+(k-a)(\alpha-1))^{\underline{(k-a)\alpha-1}} \\
&= \lambda \sum_{k=a}^{t-\alpha+1} \frac{\lambda^{k-a}}{\Gamma((k+1-a)\alpha)}(t-a+(k-a)(\alpha-1))^{\underline{(k+1-a)\alpha-1}} \\
&= \lambda E(t, a, -1, a, \alpha, \alpha, \lambda) \leq 0,
\end{aligned}$$

whence $E(t, a+\alpha-1, -1, a, \alpha, 1, \lambda)$ is a decreasing function of t. By Theorem 3.6 and Corollary 3.17, we deduce the existence of a number $L \geq 0$ such that

$$\lim_{t\to\infty} E(t, a+\alpha-1, -1, a, \alpha, 1, \lambda) = L.$$

By assuming $L > 0$, we will get a contradiction. Denote by $x(t) = E(t, a+\alpha-1, -1, a, \alpha, 1, \lambda)$. By Lemma 3.1, we have, for $t = a+\alpha-1+k$,

$$x(t) = x(a+\alpha-1) + \frac{\lambda}{\Gamma(\alpha)} \sum_{s=a}^{t-\alpha} (t-(s+1))^{\underline{\alpha-1}} x(s+\alpha-1), \; t \in \mathbb{N}_{a+\alpha-1}.$$

$$= x(a+\alpha-1) + \lambda \Bigg[x(t-1) + \alpha x(t-2)$$

$$+ \frac{\alpha(\alpha+1)}{2!} x(t-3) + \cdots + \frac{\alpha \cdots (\alpha+k-3)(\alpha+k-2)}{(k-1)!} x(a+\alpha-1) \Bigg].$$

Let $k_0 \in \mathbb{N}_2$ be fixed and t sufficiently large. Since $\lambda < 0$, we have

$$x(t) \le x(a+\alpha-1) + \lambda \Bigg[x(t-1) + \alpha x(t-2)$$

$$+ \frac{\alpha(\alpha+1)}{2!} x(t-3) + \cdots + \frac{\alpha(\alpha+1)\cdots(\alpha+k_0-1)}{k_0!} x(t-k_0-1) \Bigg].$$

Letting $t \to \infty$, we get

$$0 < L < x(a+\alpha-1)$$

$$+ \lambda L \left[1 + \alpha + \frac{\alpha(\alpha+1)}{2!} + \cdots + \frac{\alpha(\alpha+1)\cdots(\alpha+k_0-1)}{k_0!} \right]. \qquad (3.17)$$

Now notice that

$$1 + \alpha + \frac{\alpha(\alpha+1)}{2!} + \cdots + \frac{\alpha(\alpha+1)\cdots(\alpha+k_0-1)}{k_0!}$$

$$= \frac{(\alpha+1)(\alpha+2)\cdots(\alpha+k_0)}{k_0!}$$

$$= \frac{(\alpha+1)(\alpha+2)\cdots(\alpha+1+k_0-1)}{(k_0-1)!(k_0-1)^{\alpha+1}} \frac{(k_0-1)^{\alpha+1}}{k_0},$$

and using the formula (cf., for example, [35, Lemma 3.142])

$$\Gamma(z) = \lim_{n\to\infty} \frac{n! n^z}{z(z+1)\cdots(z+n)},$$

we may conclude that

$$\frac{(\alpha+1)(\alpha+2)\cdots(\alpha+1+k_0-1)}{(k_0-1)!(k_0-1)^{\alpha+1}} \frac{(k_0-1)^{\alpha+1}}{k_0} \to \infty$$

as $k_0 \to \infty$. Hence, we may choose $k_0 \in \mathbb{N}_2$ such that (3.17) is less than $-x(a + \alpha - 1)$, therefore, achieving $0 < L < 0$, which is a contradiction. The theorem is now proved.

Analogously to the proof of Theorem 3.19, we may prove the following result.

Theorem 3.23 *Assume that* $0 < \alpha < 1$ *and* $0 > \lambda \geq c(t) \geq -\alpha$ *for all* $t \in \mathbb{N}_{a+\alpha-1}$. *Then the solution of*

$$_*\Delta^{\alpha}_{a+\alpha-1}x(t) = c(t+\alpha-1)x(t+\alpha-1), \quad t \in \mathbb{N}_a$$

with $x(a+\alpha-1) \geq 0$ *satisfies*

$$x(t) \leq 2x(a+\alpha-1)E(t, a+\alpha-1, -1, a, \alpha, 1, \lambda), \; t \in \mathbb{N}_{a+\alpha-1}.$$

Finally, the following theorem follows from Theorems 3.22 and 3.24.

Theorem 3.24 *Assume* $0 < \alpha < 1$ *and* $0 > \lambda \geq c(t) \geq -\alpha$ *for all* $t \in \mathbb{N}_{a+\alpha-1}$. *Then the solution of*

$$_*\Delta^{\alpha}_{a+\alpha-1}x(t) = c(t+\alpha-1)x(t+\alpha-1), \quad t \in \mathbb{N}_a$$

with $x(a+\alpha-1) > 0$ *satisfies* $\lim_{t\to\infty} x(t) = 0$.

In all of the above analysis we considered $x(a+\alpha-1) \geq 0$. We may consider $x(a+\alpha-1) \leq 0$ and still deduce analogous results. More precisely, by making the transformation $x(t) = -y(t)$ and using Theorems 3.21 and 3.24, we get the following results, respectively.

Theorem 3.25 *Assume* $0 < \alpha < 1$ *and* $c(t) \geq \lambda > 0$ *for all* $t \in \mathbb{N}_{a+\alpha-1}$. *Then the solution of*

$$_*\Delta^{\alpha}_{a+\alpha-1}x(t) = c(t+\alpha-1)x(t+\alpha-1), \quad t \in \mathbb{N}_a$$

with $x(a+\alpha-1) < 0$ *satisfies* $\lim_{t\to\infty} x(t) = -\infty$.

Theorem 3.26 *Assume* $0 < \alpha < 1$ *and* $0 > \lambda \geq c(t) \geq -\alpha$ *for all* $t \in \mathbb{N}_{a+\alpha-1}$. *Then the solution of*

$$_*\Delta^{\alpha}_{a+\alpha-1}x(t) = c(t+\alpha-1)x(t+\alpha-1), \quad t \in \mathbb{N}_a$$

with $x(a+\alpha-1) < 0$ *satisfies* $\lim_{t\to\infty} x(t) = 0$.

3.1.3 Stability*

In this section we will present the main results obtained in [16], which essentially provide criteria regarding the stability of the fractional difference system

$$ {}_*\Delta^{\alpha}_{\alpha-1}x(t) = Mx(t+\alpha-1), \quad t \in \mathbb{N}_0,\ 0<\alpha\le 1, \tag{3.18}$$

$$ x(\alpha-1) = x_{\alpha-1}, \tag{3.19}$$

where, here, M is a square matrix of dimension $d \in \mathbb{N}_1$ and $x_{\alpha-1}$ is a vector of the same dimension.

We start by defining the *stability concepts* related to the problem above (within this section, $\|\cdot\|$ stands for an appropriate norm on $\mathbb{R}^d$).

Definition 3.27 ([16, Definition 2.1]) The system (3.18) is said to be

(i) stable, if for any $x(\alpha-1) = x_{\alpha-1} \in \mathbb{R}^d$ there exists $K > 0$ such that the solution x of (3.18)–(3.19) satisfies $\|x(t)\| \le K$ for all $t \in \mathbb{N}_{\alpha-1}$.
(ii) asymptotically stable, if for any $x(\alpha-1) = x_{\alpha-1} \in \mathbb{R}^d$ the solution x of (3.18)–(3.19) satisfies $\|x(t)\| \to 0$ as $t \to \infty$ within the domain $\mathbb{N}_{\alpha-1}$.

Let us introduce the set

$$S^{\alpha} = \left\{ z \in \mathbb{C} : |z| < \left(2\cos\frac{|\arg z| - \pi}{2-\alpha} \right)^{\alpha} \text{ and } |\arg z| > \frac{\alpha\pi}{2} \right\}.$$

It follows the main result in [16].

Theorem 3.28 ([16, Theorem 1.4]) *Let* $0 < \alpha < 1$ *and* $M \in \mathcal{M}_d$ *(the space of square matrices of dimension* d*). If* $\mu \in S^{\alpha}$ *for all the eigenvalues* μ *of* M*, then* (3.18) *is asymptotically stable. In this case, the solutions of* (3.18) *decay toward zero algebraically (and not exponentially), more precisely*

$$\|x(t)\| = O(t^{-\alpha}) \text{ as } t \to \infty.$$

Furthermore, if $\mu \in \mathbb{C}\backslash cl(S^{\alpha})$ *for an eigenvalue* μ *of* M*, then* (3.18) *is not stable.*

Remark 3.29 The stability boundary of S^{α} can be described via parametric equations

$$\operatorname{Re} z = -2(\cos\theta)^{\alpha}\cos(2-\alpha)\theta,$$

$$\operatorname{Im} z = -2(\cos\theta)^{\alpha}\sin(2-\alpha)\theta,$$

where $|\theta| \le \pi/2$. We may observe that this curve is reduced to the circle $(\operatorname{Re} z + 1)^2 + (\operatorname{Im} z)^2 = 1$ when $\alpha = 1$. In this case, Theorem 3.28 is essentially [19, Theorem 4.13].

Now we would like to mention an immediate consequence of Theorem 3.28, which illustrates that, investigating the planar system (3.18), the relevant stability conditions can be reformulated directly in terms of the matrix coefficients. To simplify the notation, we put

$$\kappa = \frac{\operatorname{tr} M}{\sqrt{4 \det M - (\operatorname{tr} M)^2)}}.$$

Corollary 3.30 *Let* $0 < \alpha < 1$ *and* $M \in \mathcal{M}_2$. *Then,* (3.18) *is asymptotically stable if* $\det M > 0$ *and either*

$$-\frac{\operatorname{tr} M}{2} \geq \sqrt{\det M}, \ \alpha > \log_2 \frac{\sqrt{(\operatorname{tr} M)^2 - 4 \det M} - \operatorname{tr} M}{2}, \tag{3.20}$$

or

$$\frac{|\operatorname{tr} M|}{2} < \sqrt{\det M} < \left(2 \cos \frac{\operatorname{arccot} \kappa - \pi}{2 - \alpha}\right)^{\alpha}, \ \alpha < \frac{2}{\pi} \operatorname{arccot} \kappa. \tag{3.21}$$

Proof The eigenvalues $\lambda_{1,2}$ of M are the zeros of the characteristic polynomial

$$\lambda^2 - \operatorname{tr} M \lambda + \det M.$$

Suppose that $\det M > 0$. We distinguish two cases:

1. Let $(\operatorname{tr} M)^2 \geq 4 \det M$. Then,

$$\lambda_{1,2} = \frac{1}{2}(\operatorname{tr} M \pm \sqrt{(\operatorname{tr} M)^2 - 4 \det M}),$$

and $\lambda_{1,2} \in S^\alpha$ if and only if

$$-2^{\alpha+1} < \operatorname{tr} M \pm \sqrt{(\operatorname{tr} M)^2 - 4 \det M} < 0. \tag{3.22}$$

So, assume $\operatorname{tr} M < 0$ in order to fulfill the previous inequalities. Then,

$$(\operatorname{tr} M)^2 \geq 4 \det M \Leftrightarrow -\frac{\operatorname{tr} M}{2} \geq \sqrt{\det M},$$

and (3.22) is equivalent to

$$\sqrt{(\operatorname{tr} M)^2 - 4 \det M} < \min\{-\operatorname{tr} M, \operatorname{tr} M + 2^{\alpha+1}\},$$

which yields (3.20). The result now follows from Theorem 3.28.

2. Let $(\operatorname{tr} M)^2 < 4 \det M$. Then,

$$\lambda_{1,2} = \frac{1}{2}(\operatorname{tr} M \pm \mathrm{i}\sqrt{4 \det M - (\operatorname{tr} M)^2}).$$

Because of symmetry of the stability area S^α along the real axis, it is enough to discuss the location of λ_1 with respect to S^α. Since

$$|\lambda_1| = \sqrt{\det M} \text{ and } \arg \lambda_1 = \operatorname{arccot} \kappa,$$

the stability condition of Theorem 3.28 becomes

$$\sqrt{\det M} < \left(2 \cos \frac{\operatorname{arccot} \kappa - \pi}{2 - \alpha}\right)^\alpha \text{ and } \operatorname{arccot} \kappa > \frac{\alpha\pi}{2},$$

which is equivalent to (3.21).

The proof is done.

Example 3.31 Consider the following system of fractional difference equations:

$$_*\Delta^\alpha_{\alpha-1} x_1(t) = x_2(t + \alpha - 1), \tag{3.23}$$

$$_*\Delta^\alpha_{\alpha-1} x_2(t) = -kx_1(t + \alpha - 1) - cx_2(t + \alpha - 1), \tag{3.24}$$

where $t \in \mathbb{N}_0$ and c, k are real constants. Then, putting

$$M = \begin{pmatrix} 0 & 1 \\ -k & -c \end{pmatrix},$$

we get $\operatorname{tr} M = -c$ and $\det M = k$. Suppose that $k > 0$. Then, the stability region (3.20) is

$$\frac{c}{2} \geq \sqrt{k}, \quad \alpha > \log_2 \frac{\sqrt{c^2 - 4k} + c}{2}.$$

Observe that the two inequalities above imply that $k < 2^{2\alpha}$. Therefore we conclude, after some calculations, that the fractional difference system (3.23)–(3.24) is asymptotically stable if

$$2\sqrt{k} \leq c < 2^{-\alpha}k + 2^\alpha, \quad 0 < k < 2^{2\alpha}.$$

Regarding the stability region (3.21), it can be shown to be equivalent to (cf. [16, pag. 669]) (see also Fig. 3.1)

$$2\sqrt{k} \cos\left((2 - \alpha) \arccos \frac{k^{1/2\alpha}}{2}\right)^\alpha < c < 2\sqrt{k} \cos \frac{\alpha\pi}{2}. \tag{3.25}$$

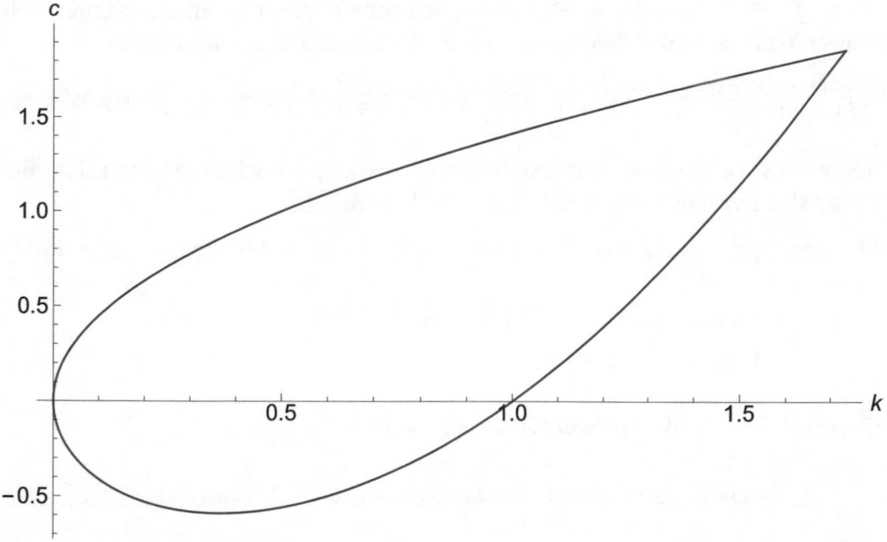

Fig. 3.1 Asymptotic stability area for (3.25) depicted in the (k, c)-plane when $\alpha = \frac{1}{2}$

It follows that the fractional difference system (3.23)–(3.24) is also asymptotically stable in this case.

3.2 Nonlinear Difference Equations: Influence of Perturbed Data

We have seen at the beginning of this chapter that the DFIVP (3.1)–(3.2) has a unique solution (cf. Remark 3.2). In this section we wish to study how *small* changes in the data influence the solution of the DFIVP. We recall that, traditionally, a problem is called *well-posed* if it has the following three properties:

- A solution exists.
- The solution is unique.
- The solution depends on the given data in a continuous way.

In what follows we will be concerned with the third item above. In our first result in this direction, we investigate the dependence of the solution of a fractional difference equation on the initial values.

So, consider the following DFIVP:

$$ {}_*\Delta^{\alpha}_{a+\alpha-1} x(t) = f(t+\alpha-1, x(t+\alpha-1)), \quad t \in \mathbb{N}_a, \tag{3.26} $$

$$ x(a+\alpha-1) = A, \tag{3.27} $$

where $f : \mathbb{N}_{a+\alpha-1} \times \mathbb{R} \to \mathbb{R}$ is a function satisfying a Lipschitz condition with respect to the second variable, i.e., there exists a constant $L > 0$ such that

$$|f(t, x_1) - f(t, x_2)| \leq L|x_1 - x_2|, \quad (t, x_i) \in \mathbb{N}_{a+\alpha-1} \times \mathbb{R}, \ i \in \{1, 2\}. \tag{3.28}$$

We now prove the first main result of this section, for which we appeal to the Gronwall's inequality obtained before (cf. Theorem 3.9).

Theorem 3.32 *Let x be the solution of* (3.26)–(3.27), *and let y be the solution of*

$$\begin{aligned} {}_*\Delta^{\alpha}_{a+\alpha-1} y(t) &= f(t + \alpha - 1, y(t + \alpha - 1)), \quad t \in \mathbb{N}_a, \\ y(a + \alpha - 1) &= B. \end{aligned}$$

Moreover, let $\varepsilon = |A - B|$ and $\delta(t) = x(t) - y(t)$. Then,

$$|\delta(t)| \leq \varepsilon E(t, a + \alpha - 1, -1, a, \alpha, 1, L), \ t \in \mathbb{N}_{a+\alpha-1}.$$

Proof Let $\delta(t) = x(t) - y(t)$. Then, the function δ solves the following DIVP:

$$\begin{aligned} {}_*\Delta^{\alpha}_{a+\alpha-1} \delta(t) &= f(t + \alpha - 1, x(t + \alpha - 1)) - f(t + \alpha - 1, y(t + \alpha - 1)), \\ \delta(a + \alpha - 1) &= A - B. \end{aligned}$$

Using Lemma 3.1, we may write

$$\begin{aligned} \delta(t) &= A - B \\ &\quad + \frac{1}{\Gamma(\alpha)} \sum_{s=a}^{t-\alpha} (t - (s + 1))^{\underline{\alpha-1}} \\ &\quad \times [f(s + \alpha - 1, x(s + \alpha - 1)) - f(s + \alpha - 1, y(s + \alpha - 1))], \end{aligned}$$

for $t \in \mathbb{N}_{a+\alpha-1}$. Using the Lipschitz condition (3.28), we obtain

$$|\delta(t)| \leq |A - B| + \Delta_a^{-\alpha}[L|\delta(s + \alpha - 1)|](t), \quad L > 0,$$

and from Remark 3.10,

$$|\delta(t)| \leq |A - B| E(t, a + \alpha - 1, -1, a, \alpha, 1, L), \ t \in \mathbb{N}_{a+\alpha-1},$$

which concludes the proof.

Remark 3.33 Theorem 3.32 essentially tells us that, as long as we do not change too much the initial condition (3.27) of the DFIVP then, on a closed interval $\{a + \alpha, a + \alpha + 1, \ldots, b\}$ with $b \in \mathbb{N}_{a+\alpha}$, the two solutions are close enough.

Next we look at the influence of changes in the given function on the right-hand side of the fractional difference equation.

Theorem 3.34 *Let x be the solution of* (3.26)–(3.27), *and let y be the solution of*

$$\begin{aligned} {}_*\Delta^{\alpha}_{a+\alpha-1}y(t) &= \tilde{f}(t+\alpha-1, y(t+\alpha-1)), \quad t \in \mathbb{N}_a, \\ y(a+\alpha-1) &= A, \end{aligned}$$

where $\tilde{f}$ is a small perturbation of f, in the sense that, we assume the existence of a number $\varepsilon > 0$ such that $|f(t,x) - \tilde{f}(t,x)| \le \varepsilon$ for all $(t,x) \in \mathbb{N}_{a+\alpha-1} \times \mathbb{R}$.

Let $\delta(t) = x(t) - y(t)$. Then, for $b = a + k$ with $k \in \mathbb{N}_0$, we have

$$|\delta(t)| \le \varepsilon \frac{(b+\alpha-a)^{\underline{\alpha}}}{\Gamma(\alpha+1)} E(t, a+\alpha-1, -1, a, \alpha, 1, L), \; t \in \mathbb{N}^{b+\alpha}_{a+\alpha-1}.$$

Proof Let $\delta(t) = x(t) - y(t)$. Then, the function δ solves the following DIVP:

$$\begin{aligned} {}_*\Delta^{\alpha}_{a+\alpha-1}\delta(t) &= f(t+\alpha-1, x(t+\alpha-1)) - \tilde{f}(t+\alpha-1, y(t+\alpha-1)), \\ \delta(a+\alpha-1) &= 0. \end{aligned}$$

Using Lemma 3.1, we may write

$$\begin{aligned} \delta(t) = &\frac{1}{\Gamma(\alpha)} \sum_{s=a}^{t-\alpha} (t-(s+1))^{\underline{\alpha-1}} \\ &\times [f(s+\alpha-1, x(s+\alpha-1)) - \tilde{f}(s+\alpha-1, y(s+\alpha-1))], \end{aligned}$$

for $t \in \mathbb{N}_{a+\alpha-1}$. Using the Lipschitz condition (3.28) and also (2.10), we obtain

$$\begin{aligned} |\delta(t)| \le &\frac{1}{\Gamma(\alpha)} \sum_{s=a}^{t-\alpha} (t-(s+1))^{\underline{\alpha-1}} \cdot \Big(|f(s+\alpha-1, x(s+\alpha-1)) \\ &- f(s+\alpha-1, y(s+\alpha-1))| \\ &+ |f(s+\alpha-1, y(s+\alpha-1)) - \tilde{f}(s+\alpha-1, y(s+\alpha-1))| \Big) \\ \le &\, L\Delta_a^{-\alpha}[|\delta(s+\alpha-1)|](t) + \varepsilon \frac{(b+\alpha-a)^{\underline{\alpha}}}{\Gamma(\alpha+1)}, \end{aligned}$$

and from Remark 3.10,

$$|\delta(t)| \le \varepsilon \frac{(b+\alpha-a)^{\underline{\alpha}}}{\Gamma(\alpha+1)} E(t, a+\alpha-1, -1, a, \alpha, 1, L), \; t \in \mathbb{N}^{b+\alpha}_{a+\alpha-1}.$$

The proof is done.

We conclude this section by calling the attention to the reader that, in a fractional problem, we may also perturb the order of the difference equation α (this being in contrast with a classical difference equation). We do not pursuit that analysis in this text, and we refer the reader to the work by Goodrich [31] for gathering information and knowledge on this sort of study.

3.3 Boundary Value Problems

In this part of the book we move away from the initial value problems discussed so far and turn our attention to (discrete) boundary value problems (BVPs) of order $0 < \alpha \leq 1$. More specifically, we will consider the following BVP:

$$ {}_*\Delta^{\alpha}_{a+\alpha-1}x(t) = f(t+\alpha-1, x(t+\alpha-1)),\ t \in \mathbb{N}_a^{a+T},\ a \in \mathbb{R},\ T \in \mathbb{N}_1, \tag{3.29} $$

$$ \delta x(a+\alpha-1) + \gamma x(a+\alpha+T) = \beta,\ \delta, \gamma, \beta \in \mathbb{R}. \tag{3.30} $$

We start by reformulating the boundary value problem (3.29)–(3.30) as a summation equation.

Lemma 3.35 *Let* $0 < \alpha \leq 1$ *and* $\delta, \beta, \gamma \in \mathbb{R}$ *be such that* $\delta + \gamma \neq 0$. *Then, for* $T \in \mathbb{N}_1$, x *is a solution of the boundary value problem* (3.29)–(3.30) *if and only if* x *satisfies the following equality:*

$$ \begin{aligned} x(t) = {} & \frac{\beta}{\delta+\gamma} + \frac{1}{\Gamma(\alpha)} \sum_{s=a}^{t-\alpha} (t-(s+1))^{\underline{\alpha-1}} f(s+\alpha-1, x(s+\alpha-1)) \\ & - \frac{\gamma}{(\delta+\gamma)\Gamma(\alpha)} \\ & \times \sum_{s=a}^{a+T} (a+\alpha+T-(s+1))^{\underline{\alpha-1}} f(s+\alpha-1, x(s+\alpha-1)),\ t \in \mathbb{N}_{a+\alpha-1}^{a+\alpha+T}. \end{aligned} $$

Proof Let us start observing that $x : \mathbb{N}_{a+\alpha-1}^{a+\alpha+T} \to \mathbb{R}$ solves (3.29) if and only if it solves the following summation equation:

$$ x(t) = c + \frac{1}{\Gamma(\alpha)} \sum_{s=a}^{t-\alpha} (t-(s+1))^{\underline{\alpha-1}} f(s+\alpha-1, x(s+\alpha-1)), \tag{3.31} $$

where c is a real constant.

Now, $x(a+\alpha-1)=c$ and

$$x(a+\alpha+T)=c+\frac{1}{\Gamma(\alpha)}\sum_{s=a}^{a+T}(a+\alpha+T-(s+1))^{\underline{\alpha-1}}f(s+\alpha-1,x(s+\alpha-1)).$$

By (3.30),

$$\delta c+\gamma$$
$$\cdot\left(c+\frac{1}{\Gamma(\alpha)}\sum_{s=a}^{a+T}(a+\alpha+T-(s+1))^{\underline{\alpha-1}}f(s+\alpha-1,x(s+\alpha-1))\right)=\beta,$$

from which follows that

$$c=\frac{1}{\delta+\gamma}$$
$$\cdot\left(\beta-\frac{\gamma}{\Gamma(\alpha)}\sum_{s=a}^{a+T}(a+\alpha+T-(s+1))^{\underline{\alpha-1}}f(s+\alpha-1,x(s+\alpha-1))\right).$$

The result now follows immediately.

Remark 3.36 Note that if we let $\gamma=0$ and $\delta=1$ in (3.29)–(3.30), then we obtain the initial value problem studied previously in this text (3.1)–(3.2).

Lemma 3.35 immediately allows us to deduce a simple existence and uniqueness theorem concerning the boundary value problem (3.29)–(3.30).

Theorem 3.37 *Assume that the function* $f:\mathbb{N}_{a+\alpha-1}^{a+T+\alpha-1}\times\mathbb{R}\to\mathbb{R}$ *is continuous and that there exists a constant* $L>0$ *such that*

$$|f(t,x)-f(t,y)|\leq L|x-y|,\quad t\in\mathbb{N}_{a+\alpha-1}^{a+T+\alpha-1},\ x,y\in\mathbb{R}. \tag{3.32}$$

If $\delta+\gamma\neq 0$ *and*

$$L\frac{(\alpha+T)^{\underline{\alpha}}}{\Gamma(\alpha+1)}\left(1+\frac{|\gamma|}{\Gamma(\alpha)|(\delta+\gamma)|}\right)<1, \tag{3.33}$$

then the boundary value problem (3.29)–(3.30) *has a unique solution defined on* $\mathbb{N}_{a+\alpha-1}^{a+T+\alpha}$.

Proof Consider the space of functions $X=\{u:\mathbb{N}_{a+\alpha-1}^{a+T+\alpha}\to\mathbb{R}\}$ endowed with norm

$$\|u\|=\sup_{t\in\mathbb{N}_{a+\alpha-1}^{a+T+\alpha}}|u(t)|.$$

Define the operator $F : X \to X$ by

$$
\begin{aligned}
&F[u](t) \\
&= \frac{\beta}{\delta+\gamma} + \frac{1}{\Gamma(\alpha)} \sum_{s=a}^{t-\alpha} (t-(s+1))^{\underline{\alpha-1}} f(s+\alpha-1, u(s+\alpha-1)) \\
&\quad - \frac{\gamma}{(\delta+\gamma)\Gamma(\alpha)} \sum_{s=a}^{a+T} (a+\alpha+T-(s+1))^{\underline{\alpha-1}} f(s+\alpha-1, u(s+\alpha-1)).
\end{aligned}
$$

By Lemma 3.35, it follows that the fixed points of the operator F, i.e., the functions $u \in X$ such that $F[u] = u$, are the solutions of the boundary value problem (3.29)–(3.30). Now, for $x, y \in X$, we have for each $t \in \mathbb{N}_{a+\alpha-1}^{a+T+\alpha}$

$$
\begin{aligned}
&|F[x](t) - F[y](t)| \\
&\le \frac{1}{\Gamma(\alpha)} \sum_{s=a}^{t-\alpha} (t-(s+1))^{\underline{\alpha-1}} \\
&\quad \times |f(s+\alpha-1, x(s+\alpha-1)) - f(s+\alpha-1, y(s+\alpha-1))| \\
&\quad + \frac{|\gamma|}{\Gamma(\alpha)|(\delta+\gamma)|} \sum_{s=a}^{a+T} (a+\alpha+T-(s+1))^{\underline{\alpha-1}} \\
&\quad \times |f(s+\alpha-1, x(s+\alpha-1)) - f(s+\alpha-1, y(s+\alpha-1))| \\
&\le L\|x-y\| \frac{1}{\Gamma(\alpha)} \sum_{s=a}^{t-\alpha} (t-(s+1))^{\underline{\alpha-1}} \\
&\quad + L\|x-y\| \frac{|\gamma|}{\Gamma(\alpha)|(\delta+\gamma)|} \sum_{s=a}^{a+T} (a+\alpha+T-(s+1))^{\underline{\alpha-1}} \\
&= L\|x-y\| \left(\frac{(t-a)^{\underline{\alpha}}}{\Gamma(\alpha+1)} + \frac{|\gamma|}{\Gamma(\alpha)|(\delta+\gamma)|} \frac{(\alpha+T)^{\underline{\alpha}}}{\Gamma(\alpha+1)} \right) \\
&\le L\|x-y\| \frac{(\alpha+T)^{\underline{\alpha}}}{\Gamma(\alpha+1)} \left(1 + \frac{|\gamma|}{\Gamma(\alpha)|(\delta+\gamma)|} \right),
\end{aligned}
$$

where we have used Corollary 2.18. Thus,

$$
\|F[x] - F[y]\| \le L \frac{(\alpha+T)^{\underline{\alpha}}}{\Gamma(\alpha+1)} \left(1 + \frac{|\gamma|}{\Gamma(\alpha)|(\delta+\gamma)|} \right) \|x-y\|,
$$

and since (3.33) holds, the result follows by the Banach fixed point theorem (cf. [42, Theorem 1.9]).

Example 3.38 Consider the following *terminal value problem* (i.e., $\delta = 0$ and $\gamma = 1$):

$$\begin{aligned}
{}_*\Delta_{-1/2}^{1/2}x(t) &= \frac{|x(t-1/2)|}{10(1+|x(t-1/2)|)}, \quad t \in \mathbb{N}_0^{10},\\
x\left(\frac{21}{2}\right) &= 1.
\end{aligned}$$

Set

$$f(t,x) = \frac{|x|}{10(1+|x|)}, \qquad t, x \in \mathbb{R}.$$

Then,

$$\begin{aligned}
|f(t,x) - f(t,y)| &= \left|\frac{|x|}{10(1+|x|)} - \frac{|y|}{10(1+|y|)}\right|\\
&= \frac{1}{10}\left|\frac{|x|-|y|}{(1+|x|)(1+|y|)}\right|\\
&\leq \frac{1}{10}||x|-|y|| \leq \frac{1}{10}|x-y|,
\end{aligned}$$

for all $t, x, y \in \mathbb{R}$. Now, for $L = 1/10$, $\alpha = 1/2$, $T = 10$, $\delta = 0$, and $\gamma = 1$, it is easy to show that (3.33) holds. Therefore, by Theorem 3.37, the terminal value problem above has a unique solution.

In the theory of difference equations of integer order, the prominent class of boundary value problems studied is the one with second-order difference equations (cf. [41, Chapter 9]). Here we will not pursuit such a goal, and we refer the reader to [8, 33, 38]) in order to find some results within the subject.

3.4 Exercises

1. Solve the following initial value problems:

   (a) $\Delta y(t) = 3^t y(t)$, $y(0) = 2$ on $\mathbb{N}_0$.
   (b) $\Delta_{-0.1}^{0.9} y(t) = (t-0.1)^{\underline{5}} y(t-0.1)$, $y(-0.1) = 0$ on $\mathbb{N}_0$.

2. Plot the solution of

$$ {}_*\Delta_{-\frac{1}{2}}^{\frac{1}{2}} y(t) = \lambda y\left(t - \frac{1}{2}\right), \quad y\left(-\frac{1}{2}\right) = 1,$$

   for $\lambda \in \{-2, 2, 0\}$.

3. Verify the following equalities:

$$T_y^{t+1-a}[1](t) = 0, \quad t \in \mathbb{N}_a,$$

$$T_y^{k-a}[1](t+1) = T_y^{k-a}[1](t) + y(t)T_y^{k-1-a}[1](t), \quad k \in \mathbb{N}_{a+1}^{t+1}.$$

4. Prove Corollary 3.17.
5. Consider the following system:

$$_*\Delta_{-\frac{3}{4}}^{\frac{1}{4}} x_1(t) = x_2\left(t - \frac{3}{4}\right), \tag{3.34}$$

$$_*\Delta_{-\frac{3}{4}}^{\frac{1}{4}} x_2(t) = -x_1\left(t - \frac{3}{4}\right) - 2x_2\left(t - \frac{3}{4}\right). \tag{3.35}$$

   Show that (3.34)–(3.35) is asymptotically stable.
6. Verify which of the following functions satisfy the inequality in (3.32):

   (a) $f(x) = \sqrt{x}$, $x \in \mathbb{R}_0^+$.
   (b) $f(x) = \sqrt{x^2 + 5}$, $x \in \mathbb{R}$.

Chapter 4
Calculus of Variations

The calculus of variations has a long and rich history of interaction with other branches of mathematics such as geometry and differential equations, and with physics, particularly mechanics. In more recent years, the theory of calculus of variations has been used by economical scientists as well as by electrical engineers. This beautiful theory is nearly as old as the calculus, and both subjects were developed somewhat in parallel. The classical continuous calculus of variations deals with finding extrema for functionals;[1] the candidates in competition for an *extremum* are thus functions as opposed to vectors in $\mathbb{R}^n$.

We aim here in this chapter to develop the foundational theory of the calculus of variations for problems in which discrete fractional operators are present (doing it in analogy with the continuous case—cf. [24, 29]). We will consider the *basic problem of the discrete fractional calculus of variations*, i.e., for $a \in \mathbb{R}$, $b = a + k$ with $k \in \mathbb{N}_2$ and $0 < \alpha \leq 1$, we wish to find functions $y : \mathbb{N}_{a+\alpha-1}^{b+\alpha-1} \to \mathbb{R}$ such that

$$\mathcal{L}(y) = \frac{1}{\Gamma(\alpha)} \sum_{t=a}^{b-1} (b + \alpha - 1 - (t+1))^{\underline{\alpha-1}} L(t, y(t+\alpha), {}_*\Delta_{a+\alpha-1}^{\alpha} y(t)) \to \min$$

$$y(a + \alpha - 1) = A, \ y(b + \alpha - 1) = B, \tag{P}$$

where $L : \mathbb{N}_a^{b-1} \times \mathbb{R}^2 \to \mathbb{R}$ and (recall) ${}_*\Delta_{a+\alpha-1}^{\alpha}$ is the Caputo fractional difference of order α.

Definition 4.1 Consider the set of admissible functions to (P), i.e., let the Lagrangean $\mathcal{L}$ be defined on the set $\mathcal{F} = \{y : \mathbb{N}_{a+\alpha-1}^{b+\alpha-1} \to \mathbb{R} : y(a + \alpha - 1) = A,\ y(b + \alpha - 1) = B\}$.

[1] A *functional* is a mapping from a set of functions to the real numbers.

R. A. C. Ferreira, *Discrete Fractional Calculus and Fractional Difference Equations*, SpringerBriefs in Mathematics,
https://doi.org/10.1007/978-3-030-92724-0_4

We say that $\tilde{y} \in \mathcal{F}$ solves (P) locally if there exists $\delta > 0$ such that $\mathcal{L}(\tilde{y}) \leq \mathcal{L}(y)$ for all $y \in \mathcal{F}$ with $\|y - \tilde{y}\| < \delta$, where $\|y - \tilde{y}\| = \max_{t\in\{a,\ldots,b\}} |y(t+\alpha-1) - \tilde{y}(t+\alpha-1)|$.

We say that $\tilde{y} \in \mathcal{F}$ solves (P) globally if $\mathcal{L}(\tilde{y}) \leq \mathcal{L}(y)$ for all $y \in \mathcal{F}$.

Remark 4.2 We remind the reader that there is no loss of generality in considering a minimization problem in (P) as, to consider the maximization problem, one can apply all the subsequent theory to the functional $-\mathcal{L}$.

Definition 4.3 A function $\eta : \mathbb{N}_{a+\alpha-1}^{b+\alpha-1} \to \mathbb{R}$ is called an admissible variation if $\eta(a+\alpha-1) = \eta(b+\alpha-1) = 0$. The set of admissible variations is denoted by $\mathcal{V}$.

Lemma 4.4 (Fundamental Lemma of the Discrete Calculus of Variations) *Suppose that* $f : \mathbb{N}_a^{b-1} \to \mathbb{R}$. *Then, for* $0 < \alpha \leq 1$,

$$\sum_{t=a}^{b-1} f(t)\eta(t+\alpha) = 0, \ \forall \eta \in \mathcal{V}, \tag{4.1}$$

if and only if $f(t) = 0$ *for all* $t \in \mathbb{N}_a^{b-2}$.

Proof If $f(t) = 0$ for all $t \in \mathbb{N}_a^{b-2}$, the result is obvious.

Suppose now that $f(t_0) \neq 0$ for some $t_0 \in \mathbb{N}_a^{b-2}$. Define $\eta(t_0+\alpha) = f(t_0)$ and $\eta(t) = 0$ at the other points t of $\mathbb{N}_{a+\alpha-1}^{b+\alpha-1}$. Then, (4.1) does not hold, and this concludes the proof.

For completeness, we also present a discrete version of the du Bois-Reymond lemma.

Lemma 4.5 (Discrete Fractional du Bois-Reymond Lemma) *Suppose that* $f : \mathbb{N}_a^{b-1} \to \mathbb{R}$. *Then,*

$$\frac{1}{\Gamma(\alpha)} \sum_{t=a}^{b-1} (b+\alpha-1-(s+1))^{\underline{\alpha-1}} f(t)_* \Delta_{a+\alpha-1}^{\alpha} \eta(t) = 0, \ \forall \eta \in \mathcal{V}, \tag{4.2}$$

if and only if $f(t) = k \in \mathbb{R}$ *for all* $t \in \mathbb{N}_a^{b-1}$.

Proof We start by assuming that $f(t) = k$ on $\mathbb{N}_a^{b-1}$. Then, for $\eta \in \mathcal{V}$, we have

$$\begin{aligned} &\frac{1}{\Gamma(\alpha)} \sum_{t=a}^{b-1} (b+\alpha-1-(s+1))^{\underline{\alpha-1}} k_* \Delta_{a+\alpha-1}^{\alpha} \eta(t) \\ &\quad = k \Delta_a^{-\alpha}{}_* \Delta_{a+\alpha-1}^{\alpha} \eta(b+\alpha-1) = \eta(b+\alpha-1) - \eta(a+\alpha-1) = 0, \end{aligned}$$

where we have used Theorem 2.26.

Suppose now that (4.2) holds with $f : \mathbb{N}_a^{b-1} \to \mathbb{R}$. Define $\eta : \mathbb{N}_{a+\alpha-1}^{b+\alpha-1} \to \mathbb{R}$ by

$$\eta(t) = \Delta_a^{-\alpha} f(t) - k\frac{(t-a)^{\underline{\alpha}}}{\Gamma(\alpha+1)}, \; k = \Delta_a^{-\alpha} f(b+\alpha-1)\frac{\Gamma(\alpha+1)}{(b+\alpha-1-a)^{\underline{\alpha}}}.$$

It is clear that $\eta(a+\alpha-1) = \eta(b+\alpha-1) = 0$, hence $\eta \in \mathcal{V}$. Moreover, ${}_*\Delta_{a+\alpha-1}^{\alpha}\eta(t) = f(t) - k$ by Theorem 2.24 and (2.13). We then have

$$\begin{aligned} 0 &= \frac{1}{\Gamma(\alpha)}\sum_{t=a}^{b-1}(b+\alpha-1-(s+1))^{\underline{\alpha-1}}(f(t)-k+k)_*\Delta_{a+\alpha-1}^{\alpha}\eta(t) \\ &= \frac{1}{\Gamma(\alpha)}\sum_{t=a}^{b-1}(b+\alpha-1-(s+1))^{\underline{\alpha-1}}(f(t)-k)^2 + k\underbrace{\Delta_a^{-\alpha}{}_*\Delta_{a+\alpha-1}^{\alpha}\eta(b+\alpha-1)}_{=0}, \end{aligned}$$

therefore, $f(t) = k$ for all $t \in \mathbb{N}_a^{b-1}$. The proof is done.

4.1 Necessary Conditions

In this section we deduce first- and second-order necessary optimality conditions for the basic variational problem (P) to have a local *extremum*. We start with a condition of first order.

Theorem 4.6 (Euler–Lagrange Equation) *Suppose that L has continuous partial derivatives with respect to the second and third variables (denoted by L_u and L_v, respectively). Assume that $\tilde{y} \in \mathcal{F}$ solves (P) locally. Then,*

$$(b+\alpha-2-t)^{\underline{\alpha-1}}\mathbb{L}_u^{\tilde{y}}(t) + {}_{b-1}\Delta^{\alpha}[(b+\alpha-2-s)^{\underline{\alpha-1}}\mathbb{L}_v^{\tilde{y}}(s)](t+\alpha-1) = 0, \qquad (4.3)$$

for all $t \in \mathbb{N}_a^{b-2}$, where $\mathbb{L}_l^{\tilde{y}} : \mathbb{N}_a^{b-1} \to \mathbb{R}$ is the function defined by $\mathbb{L}_l^{\tilde{y}}(s) = L_l(s, \tilde{y}(s+\alpha), {}_*\Delta_{a+\alpha-1}^{\alpha}\tilde{y}(s))$.

Proof Suppose that $\tilde{y} \in \mathcal{F}$ solves (P) locally and consider an arbitrary but fixed variation $\eta \in \mathcal{V}$. For the number $\delta > 0$ prescribed in Definition 4.1, we consider the following function $\phi : \left(-\frac{\delta}{\|\eta\|}, \frac{\delta}{\|\eta\|}\right) \to \mathbb{R}$ defined by

$$\phi(\varepsilon) = \mathcal{L}(\tilde{y} + \varepsilon\eta). \qquad (4.4)$$

This function has a local minimum at $\varepsilon = 0$, so we must have $\phi'(0) = 0$, i.e.,

$$\frac{1}{\Gamma(\alpha)}\sum_{t=a}^{b-1}(b+\alpha-1-(t+1))^{\underline{\alpha-1}}\left[\mathbb{L}_u^{\tilde{y}}(t)\eta(t+\alpha) + \mathbb{L}_v^{\tilde{y}}(t)_*\Delta_{a+\alpha-1}^{\alpha}\eta(t)\right] = 0.$$

Now, using the summation by parts formula (2.7) and recalling that $\eta(a+\alpha-1) = \eta(b+\alpha-1) = 0$, we get

$$\sum_{t=a}^{b-2} \left\{ f(t)\mathbb{L}_u^{\tilde{y}}(t) + {}_{b-1}\Delta^{\alpha}[f(s)\mathbb{L}_v^{\tilde{y}}(s)](t+\alpha-1) \right\} \eta(t+\alpha) = 0,$$

where $f(t) = (b+\alpha-2-t)^{\underline{\alpha-1}}$. The conclusion now follows from Lemma 4.4.

By letting $\alpha = 1$ in the previous result, we obtain the classic discrete Euler–Lagrange equation.

Corollary 4.7 *If $\tilde{y} \in \mathcal{F}$ is a local solution of the following problem:*[2]

$$\mathcal{L}(y) = \sum_{t=a}^{b-1} L(t, y(t+1), \Delta y(t)) \to \min$$

$$y(a) = A,\ y(b) = B,$$

then

$$L_u(t, \tilde{y}(t+1), \Delta\tilde{y}(t)) - \Delta L_v(t, \tilde{y}(t+1), \Delta\tilde{y}(t)) = 0, \quad t \in \mathbb{N}_a^{b-2}.$$

Theorem 4.8 (Legendre's Condition) *Suppose that the partial derivatives L_{uu}, L_{uv} and L_{vv} are continuous. If $\tilde{y} \in \mathcal{F}$ solves* (P) *locally, then*

$$\begin{aligned}
&(b+\alpha-1-(t+1))^{\underline{\alpha-1}}[\mathbb{L}_{uu}^{\tilde{y}}(t) + 2\mathbb{L}_{uv}^{\tilde{y}}(t) + \mathbb{L}_{vv}^{\tilde{y}}(t)] \\
&\quad + (b+\alpha-1-(t+2))^{\underline{\alpha-1}}\alpha^2\mathbb{L}_{vv}^{\tilde{y}}(t+1) \\
&\quad + \sum_{s=t+2}^{b-1} (b+\alpha-1-(s+1))^{\underline{\alpha-1}}\mathbb{L}_{vv}^{\tilde{y}}(s)\binom{s-1-(t+\alpha)}{-\alpha-1}^2 \geq 0,
\end{aligned}$$

for all $t \in \mathbb{N}_a^{b-2}$, where $\mathbb{L}_{ij}^{\tilde{y}} : \mathbb{N}_a^{b-1} \to \mathbb{R}$ is the function defined by $\mathbb{L}_{ij}^{\tilde{y}}(s) = L_{ij}(s, \tilde{y}(s+\alpha), {}_*\Delta_{a+\alpha-1}^{\alpha}\tilde{y}(s))$.

Proof Suppose that $\tilde{y} \in \mathcal{F}$ solves (P) locally, and let η be an admissible variation. Fix $0 < \alpha \leq 1$ and consider, as before, the function ϕ defined by (4.4). Then,

$$\phi''(\varepsilon) \geq 0, \tag{4.5}$$

[2] Our formulation coincides with those in [12] and [27] for the time scale $\mathbb{T} = \mathbb{Z}$.

which is equivalent to

$$\frac{1}{\Gamma(\alpha)}\sum_{t=a}^{b-1}(b+\alpha-1-(t+1))^{\underline{\alpha-1}}\Big[\mathbb{L}_{uu}^{\tilde{y}}(t)\eta^2(t+\alpha)$$

$$+2\mathbb{L}_{uv}^{\tilde{y}}(t)\eta(t+\alpha)_*\Delta_{a+\alpha-1}^{\alpha}\eta(t)+\mathbb{L}_{vv}^{\tilde{y}}(t)({}_*\Delta_{a+\alpha-1}^{\alpha}\eta(t))^2\Big]\geq 0. \tag{4.6}$$

Let $\tau\in\mathbb{N}_a^{b-2}$ be arbitrary and define $\eta:\mathbb{N}_{a+\alpha-1}^{b+\alpha-1}\to\mathbb{R}$ by

$$\eta(t)=\begin{cases}1, & \text{if } t=\tau+\alpha,\\ 0, & \text{otherwise}.\end{cases}$$

It is clear that η is an admissible variation. Now note that, for $t\in\mathbb{N}_\tau^{b-1}$ and $0<\alpha<1$, we have

$$\begin{aligned}
&{}_*\Delta_{a+\alpha-1}^{\alpha}\eta(t)\\
&=\frac{1}{\Gamma(1-\alpha)}\sum_{s=a+\alpha-1}^{t+\alpha-1}(t-(s+1))^{\underline{-\alpha}}\Delta\eta(s)\\
&=\frac{\left(\sum_{s=a+\alpha-1}^{\tau+\alpha-1}(t-(s+1))^{\underline{-\alpha}}\Delta\eta(s)+\sum_{s=\tau+\alpha}^{t+\alpha-1}(t-(s+1))^{\underline{-\alpha}}\Delta\eta(s)\right)}{\Gamma(1-\alpha)}\\
&=\frac{(t-(\tau+\alpha))^{\underline{-\alpha}}-(t-(\tau+\alpha+1))^{\underline{-\alpha}}}{\Gamma(1-\alpha)}=\frac{(t-1-(\tau+\alpha))^{\underline{-\alpha-1}}}{\Gamma(-\alpha)},
\end{aligned}$$

where the last equality follows from (1.2). Notice also that when $\alpha=1$, then ${}_*\Delta_{a+\alpha-1}^{\alpha}\eta(t)=\Delta\eta(t)=\eta(t+1)-\eta(t)$ and, therefore, $\Delta\eta(t)=0$ for all $t\in\mathbb{N}_{\tau+2}^{b-1}$. Since

$${}_*\Delta_{a+\alpha-1}^{\alpha}\eta(\tau)=\Delta\eta(\tau)=1,\quad 0<\alpha<1,$$

and

$${}_*\Delta_{a+\alpha-1}^{\alpha}\eta(\tau+1)=-\alpha,\quad 0<\alpha<1,$$

we obtain from (4.6) and all $0<\alpha\leq 1$,

$$\begin{aligned}
&(b+\alpha-1-(\tau+1))^{\underline{\alpha-1}}[\mathbb{L}_{uu}^{\tilde{y}}(\tau)+2\mathbb{L}_{uv}^{\tilde{y}}(\tau)+\mathbb{L}_{vv}^{\tilde{y}}(\tau)]\\
&+(b+\alpha-1-(\tau+2))^{\underline{\alpha-1}}\alpha^2\mathbb{L}_{vv}^{\tilde{y}}(\tau+1)\\
&+\sum_{t=\tau+2}^{b-1}(b+\alpha-1-(t+1))^{\underline{\alpha-1}}\mathbb{L}_{vv}^{\tilde{y}}(t)\binom{t-1-(\tau+\alpha)}{-\alpha-1}^2\geq 0,
\end{aligned}\tag{4.7}$$

where we are using the definition in (1.6). The result is now a consequence of (4.7) and the fact that τ is arbitrary.

Corollary 4.9 *If $\tilde{y} \in \mathcal{F}$ solves locally the following problem:*

$$\mathcal{L}(y) = \sum_{t=a}^{b-1} L(t, y(t+1), \Delta y(t)) \to \min$$

$$y(a) = A,\ y(b) = B,$$

then

$$\mathbb{L}_{uu}^{\tilde{y}}(t) + 2\mathbb{L}_{uv}^{\tilde{y}}(t) + \mathbb{L}_{vv}^{\tilde{y}}(t) + \mathbb{L}_{vv}^{\tilde{y}}(t+1) \geq 0,$$

for all $t \in \mathbb{N}_a^{b-2}$, where $\mathbb{L}_{ij}^{\tilde{y}}(t) = L_{ij}(t, \tilde{y}(t+1), \Delta\tilde{y}(t))$.

Proof Just let $\alpha = 1$ in Theorem 4.8 and recall that, by (1.6), we have $\binom{s-1-(t+1)}{-2} = 0$ for $s \in \mathbb{N}_{t+2}^{b-1}$.

Example 4.10 Consider the following minimization problem:

$$\mathcal{L}(y) = \frac{1}{\Gamma(\alpha)} \sum_{t=a}^{b-1} (b+\alpha-1-(t+1))^{\underline{\alpha-1}} ({}_*\Delta_{a+\alpha-1}^{\alpha} y(t))^2 \to \min \tag{4.8}$$

$$y(a+\alpha-1) = A,\ y(b+\alpha-1) = B. \tag{4.9}$$

Let $L(t, u, v) = v^2$. Then, $L_u = 0$ and $L_v = 2v$. Suppose now that $\tilde{y}$ is a local solution of the problem given by (4.8)–(4.9). Then, the Euler–Lagrange equation reads as

$${}_{b-1}\Delta^{\alpha}[(b+\alpha-2-s)^{\underline{\alpha-1}} {}_*\Delta_{a+\alpha-1}^{\alpha} \tilde{y}(s)](t+\alpha-1) = 0,\ t \in \mathbb{N}_a^{b-2}.$$

Observe that the previous equation is equivalent to

$${}_{b-\alpha}\Delta^{\alpha}[(b-1-s)^{\underline{\alpha-1}} {}_*\Delta_{a+\alpha-1}^{\alpha} \tilde{y}(s+\alpha-1)](t) = 0,\ t \in \mathbb{N}_a^{b-2}. \tag{4.10}$$

Applying ${}_{b-2}\Delta^{-\alpha}$ to both sides of (4.10) and using Theorem 2.26, we achieve

$$(b-2-t)^{\underline{\alpha-1}} {}_*\Delta_{a+\alpha-1}^{\alpha} \tilde{y}(t+\alpha) = c(b-2-t)^{\underline{\alpha-1}},\ t \in \mathbb{N}_{a-\alpha}^{b-2-\alpha}, \tag{4.11}$$

where $c \in \mathbb{R}$.

Again, using Theorem 2.26, we obtain

$$\tilde{y}(t) = A + c(t-a)^{\underline{\alpha}}, \quad t \in \mathbb{N}_{a+\alpha-1}^{b+\alpha-1}.$$

Finally, since $y(b+\alpha-1) = B$, then $c = \frac{(B-A)\Gamma(\alpha+1)}{(b+\alpha-1-a)^{\underline{\alpha}}}$. It follows that

$$\tilde{y}(t) = A + \frac{B-A}{(b+\alpha-1-a)^{\underline{\alpha}}}(t-a)^{\underline{\alpha}}, \quad t \in \mathbb{N}_{a+\alpha-1}^{b+\alpha-1}. \tag{4.12}$$

We conclude that there is only one candidate to be the solution of (4.8)–(4.9) and is given by (4.12). We note also that the Legendre condition (cf. Theorem 4.8) is trivially satisfied by $\tilde{y}$ (in fact by any function $y : \mathbb{N}_{a+\alpha-1}^{b+\alpha-1} \to \mathbb{R}$!) because $L_{uu} = 0$, $L_{uv} = 0$, and $L_{vv} = 2 > 0$.

We will solve the minimization problem (4.8)–(4.9) hereafter in Sect. 4.3. Nevertheless we would like to point out that if $A = B$, then the constant function $\tilde{y}(t) = A$ indeed is the solution of (4.8)–(4.9), as $\mathcal{L}(y) \geq 0$ for all $y \in \mathcal{F}$ and ${}_*\Delta_{a+\alpha-1}^{\alpha}\tilde{y}(t) = 0$.

4.2 Natural Boundary Conditions

In this section we wish to make a brief detour from the basic problem of the calculus of variations (P). Concretely, we consider the problem of determining functions $y : \mathbb{N}_{a+\alpha-1}^{b+\alpha-1} \to \mathbb{R}$ such that

$$\mathcal{L}(y) = \frac{1}{\Gamma(\alpha)}\sum_{t=a}^{b-1}(b+\alpha-1-(t+1))^{\underline{\alpha-1}}L(t, y(t+\alpha), {}_*\Delta_{a+\alpha-1}^{\alpha}y(t)) \to \min. \tag{4.13}$$

The essential difference is that no boundary conditions are imposed on y. In this case, the space of admissible functions and the space of admissible variations coincide, and we denote it by $\mathbb{F} = \{f : \mathbb{N}_{a+\alpha-1}^{b+\alpha-1} \to \mathbb{R}\}$.

Now, suppose that $\tilde{y} \in \mathbb{F}$ solves (4.13) locally, and let η be an arbitrary admissible variation. Then, for the variations $\eta \in \mathcal{V}$, we may proceed as in the proof of Theorem 4.6 to conclude that $\tilde{y}$ must satisfy Euler–Lagrange equation:

$$(b+\alpha-2-t)^{\underline{\alpha-1}}\mathbb{L}_u^{\tilde{y}}(t) + {}_{b-1}\Delta^{\alpha}[(b+\alpha-2-s)^{\underline{\alpha-1}}\mathbb{L}_v^{\tilde{y}}(s)](t+\alpha-1) = 0, \tag{4.14}$$

for all $t \in \mathbb{N}_a^{b-2}$. Moreover, from the proof of Theorem 4.6, we know that for all $\eta \in \mathbb{F}$

$$\sum_{t=a}^{b-1}(b+\alpha-1-(t+1))^{\underline{\alpha-1}}\left[\mathbb{L}_u^{\tilde{y}}(t)\eta(t+\alpha) + \mathbb{L}_v^{\tilde{y}}(t){}_*\Delta_{a+\alpha-1}^{\alpha}\eta(t)\right] = 0. \tag{4.15}$$

From Theorem 2.14, we have

$$\begin{aligned}&\sum_{t=a}^{b-1}(b+\alpha-1-(t+1))^{\underline{\alpha-1}}\mathbb{L}_v^{\tilde{y}}(t)_*\Delta_{a+\alpha-1}^{\alpha}\eta(t)\\&\quad=\eta(b+\alpha-1)\Gamma(\alpha)\mathbb{L}_v^{\tilde{y}}(b-1)\\&\qquad-\eta(a+\alpha-1)_{b-1}\Delta^{-(1-\alpha)}[(b+\alpha-1-(s+1))^{\underline{\alpha-1}}\mathbb{L}_v^{\tilde{y}}(s)](a+\alpha-1)\\&\qquad+\sum_{t=a}^{b-2}\eta(t+\alpha)_{b-1}\Delta^{\alpha}[(b+\alpha-1-(s+1))^{\underline{\alpha-1}}\mathbb{L}_v^{\tilde{y}}(s)](t+\alpha-1),\end{aligned}$$

therefore, (4.15) becomes

$$\begin{aligned}&\Gamma(\alpha)\mathbb{L}_u^{\tilde{y}}(b-1)\eta(b+\alpha-1)+\eta(b+\alpha-1)\Gamma(\alpha)\mathbb{L}_v^{\tilde{y}}(b-1)\\&\quad-\eta(a+\alpha-1)_{b-1}\Delta^{-(1-\alpha)}[(b+\alpha-1-(s+1))^{\underline{\alpha-1}}\mathbb{L}_v^{\tilde{y}}(s)](a+\alpha-1)\\&\quad+\sum_{t=a}^{b-2}[f(t)\mathbb{L}_u^{\tilde{y}}(s)+{}_{b-1}\Delta^{\alpha}[f(s)\mathbb{L}_v^{\tilde{y}}(s)](t+\alpha-1)]\eta(t+\alpha)=0,\end{aligned}\tag{4.16}$$

where $f(t)=(b+\alpha-2-t)^{\underline{\alpha-1}}$. Using (4.14) in (4.16), we get

$$\begin{aligned}&\Gamma(\alpha)\eta(b+\alpha-1)[\mathbb{L}_u^{\tilde{y}}(b-1)+\mathbb{L}_v^{\tilde{y}}(b-1)]\\&\quad-\eta(a+\alpha-1)_{b-1}\Delta^{-(1-\alpha)}[(b+\alpha-1-(s+1))^{\underline{\alpha-1}}\mathbb{L}_v^{\tilde{y}}(s)](a+\alpha-1)=0.\end{aligned}\tag{4.17}$$

Now, if we use the variations for which $n(a+\alpha-1)=0$ and $n(b+\alpha-1)\neq 0$, we conclude that

$$\mathbb{L}_u^{\tilde{y}}(b-1)+\mathbb{L}_v^{\tilde{y}}(b-1)=0.$$

On the other hand, using the variations for which $n(a+\alpha-1)\neq 0$ and $n(b+\alpha-1)=0$, we obtain

$${}_{b-1}\Delta^{-(1-\alpha)}[(b+\alpha-2-s)^{\underline{\alpha-1}}\mathbb{L}_v^{\tilde{y}}(s)](a+\alpha-1)=0.$$

In summary, we have just proved the following result.

Theorem 4.11 *Suppose that $\tilde{y}\in\mathbb{F}$ solves (4.13) locally. Then, $\tilde{y}$ satisfies the Euler–Lagrange equation*

$$(b+\alpha-2-t)^{\underline{\alpha-1}}\mathbb{L}_u^{\tilde{y}}(t)+{}_{b-1}\Delta^{\alpha}[(b+\alpha-2-s)^{\underline{\alpha-1}}\mathbb{L}_v^{\tilde{y}}(s)](t+\alpha-1)=0,\ t\in\mathbb{N}_a^{b-2}$$

and the boundary conditions

$$\mathbb{L}_u^{\tilde{y}}(b-1)+\mathbb{L}_v^{\tilde{y}}(b-1)=0 \tag{4.18}$$

$$_{b-1}\Delta^{-(1-\alpha)}[(b+\alpha-2-s)^{\underline{\alpha-1}}\mathbb{L}_v^{\tilde{y}}(s)](a+\alpha-1)=0. \tag{4.19}$$

Remark 4.12 The conditions (4.18)–(4.19) are usually called *natural boundary conditions.*

Remark 4.13 We may, naturally, enunciate an analogous result to Theorem 4.11 for the problem in which only one of the boundary conditions is not present, i.e., for the minimization problem in (4.13) with $y(a)=A$ or $y(b)=B$.

Example 4.14 In order to illustrate the utility of Theorem 4.11, we consider again the functional

$$\mathcal{L}(y)=\frac{1}{\Gamma(\alpha)}\sum_{t=a}^{b-1}(b+\alpha-1-(t+1))^{\underline{\alpha-1}}({}_*\Delta_{a+\alpha-1}^{\alpha}y(t))^2,$$

and we seek function $y:\mathbb{N}_{a+\alpha-1}^{b+\alpha-1}\to\mathbb{R}$ that renders $\mathcal{L}$ a minimum. It is easily seen that any constant function is a solution to this problem, in view that $\min_y\mathcal{L}(y)=0$. Now, using Theorem 4.11, we show that they are the only solutions; for if $\tilde{y}$ is such that $\mathcal{L}(\tilde{y})=0$, then it must satisfy the Euler–Lagrange equation

$$_{b-1}\Delta^{\alpha}[(b+\alpha-2-s)^{\underline{\alpha-1}}{}_*\Delta_{a+\alpha-1}^{\alpha}\tilde{y}(s)](t+\alpha-1)=0,\ t\in\mathbb{N}_a^{b-2}$$

and the boundary conditions

$${}_*\Delta_{a+\alpha-1}^{\alpha}\tilde{y}(b-1)=0 \tag{4.20}$$

$$_{b-1}\Delta^{-(1-\alpha)}[(b+\alpha-2-s)^{\underline{\alpha-1}}{}_*\Delta_{a+\alpha-1}^{\alpha}\tilde{y}(s)](a+\alpha-1)=0. \tag{4.21}$$

We have seen before (cf. (4.11)) that

$${}_*\Delta_{a+\alpha-1}^{\alpha}\tilde{y}(t)=c,\quad t\in\mathbb{N}_a^{b-2},\ c\in\mathbb{R}.$$

Inserting this in (4.21) we obtain $c\Gamma(\alpha)=0$, which implies $c=0$. Now, recalling (4.20), we conclude that

$${}_*\Delta_{a+\alpha-1}^{\alpha}\tilde{y}(t)=0,\quad t\in\mathbb{N}_a^{b-1}.$$

From Theorem 2.26, we finally see that $\tilde{y}(t)=y(a+\alpha-1)$ for all $t\in\mathbb{N}_{a+\alpha}^{b+\alpha-1}$.

4.3 A Sufficient Condition

In this section we will provide a useful sufficient condition for the minimization Problem (P). Before proceeding, we need the following definition.

Definition 4.15 We say that a function $F : \mathbb{N}_a^b \times \mathbb{R}^2 \to \mathbb{R}$ is jointly convex in (u, v) if and only if for each $t \in \mathbb{N}_a^b$

$$F(t,u,v) - F(t,u',v') \geq (u-u')F_u(t,u',v') + (v-v')F_v(t,u',v'),\ u,v,u',v' \in \mathbb{R}, \tag{4.22}$$

provided that the partial derivatives F_u and F_v exist.

We now proceed to prove the main result of this section.

Theorem 4.16 *Suppose that the Lagrangean $L(t,u,v)$ in* (P) *is jointly convex in (u,v). Assume that the function $\tilde{y} \in \mathcal{F}$ satisfies the Euler–Lagrange equation*

$$(b+\alpha-2-t)^{\underline{\alpha-1}}\mathbb{L}_u^y(t) + {}_{b-1}\Delta^\alpha[(b+\alpha-2-s)^{\underline{\alpha-1}}\mathbb{L}_v^y(s)](t+\alpha-1) = 0, \tag{4.23}$$

for all $t \in \mathbb{N}_a^{b-2}$. Then, $\tilde{y}$ solves (P) *globally.*

Proof Throughout the proof, we will use the notation $L^y(t) = L(t, y(t+\alpha), {}_*\Delta_{a+\alpha-1}^\alpha y(t))$ besides the other ones introduced before in this chapter.

Let $y \in \mathcal{F}$ be an arbitrary function, and suppose that $\tilde{y}$ satisfies the Euler–Lagrange equation (4.23) on $t \in \mathbb{N}_a^{b-2}$. Since L is jointly convex in (u,v), we obtain from (4.22), and with the help of (2.7),

$$\begin{aligned}
\mathcal{L}(y) - \mathcal{L}(\tilde{y}) &= \frac{1}{\Gamma(\alpha)}\sum_{t=a}^{b-1}(b+\alpha-1-(t+1))^{\underline{\alpha-1}}[L^y(t) - L^{\tilde{y}}(t)] \\
&\geq \frac{1}{\Gamma(\alpha)}\sum_{t=a}^{b-1}(b+\alpha-1-(t+1))^{\underline{\alpha-1}}[(y(t+\alpha)-\tilde{y}(t+\alpha))\mathbb{L}_u^{\tilde{y}}(t) \\
&\quad + {}_*\Delta_{a+\alpha-1}^\alpha[y-\tilde{y}](t)\mathbb{L}_v^{\tilde{y}}(t)] \\
&= \frac{1}{\Gamma(\alpha)}\sum_{t=a}^{b-2}(y(t+\alpha)-\tilde{y}(t+\alpha))\{(b+\alpha-2-t)^{\underline{\alpha-1}}\mathbb{L}_u^{\tilde{y}}(t) \\
&\quad + {}_{b-1}\Delta^\alpha[(b+\alpha-2-s)^{\underline{\alpha-1}}\mathbb{L}_v^{\tilde{y}}(s)](t+\alpha-1)\},
\end{aligned}$$

where we have used the fact that $y(b-1+\alpha) - \tilde{y}(b-1+\alpha) = 0$. Finally we conclude that $\mathcal{L}(y) - \mathcal{L}(\tilde{y}) \geq 0$ from (4.23).

Example 4.17 (Example (4.10) *Revisited)* We have seen before that the function

$$\tilde{y}(t) = A + \frac{B - A}{(b + \alpha - 1 - a)^{\underline{\alpha}}}(t - a)^{\underline{\alpha}}, \quad t \in \mathbb{N}_{a+\alpha-1}^{b+\alpha-1}$$

is the unique candidate to solve the minimization problem

$$\mathcal{L}(y) = \sum_{t=a}^{b-1}(b + \alpha - 1 - (t + 1))^{\underline{\alpha - 1}}({}_*\Delta_{a+\alpha-1}^{\alpha} y(t))^2 \to \min$$

$$y(a + \alpha - 1) = A, \ y(b{+}\alpha - 1) = B.$$

We will now show that, indeed, it is its (global) solution. It was already seen that $\tilde{y}$ satisfies the Euler–Lagrange equation (4.23). Now, let $L(t, u, v) = v^2$. Then,

$$\begin{aligned} L(t, u, v) - L(t, u', v') = v^2 - v'^2 &= v^2 + v'^2 - 2vv' - 2v'^2 + 2vv' \\ &= (v - v')^2 + (v - v')2v' \geq (v - v')2v', \end{aligned}$$

i.e., L is jointly convex in (u, v). Finally, an application of Theorem 4.16 shows what was intended.

We point out that if $\alpha = 1$, we get the 'straight line' connecting the points (a, A) and (b, B) as the solution of the Euler–Lagrange equation, i.e.,

$$\tilde{y}(t) = A + \frac{B - A}{b - a}(t - a), \quad t \in \mathbb{N}_a^b.$$

4.4 Exercises

1. Find the Euler–Lagrange equation for each of the following:
   (a) $\mathcal{L}(y) = \frac{1}{\Gamma(\alpha)} \sum_{t=0}^{99}(99 + \alpha - (t + 1))^{\underline{\alpha - 1}}[4y^2(t + \alpha) + 3({}_*\Delta_{\alpha-1}^{\alpha} y(t))^2]$, with $0 < \alpha \leq 1$.
   (b) $\mathcal{L}(y) = \sum_{t=0}^{49}[y(t + \alpha) + 2(\Delta y(t))^2]$.
2. Solve the following minimization problem:

$$\mathcal{L}(y) = \frac{1}{\Gamma(\alpha)} \sum_{t=a}^{b-1}(b + \alpha - 1 - (t + 1))^{\underline{\alpha - 1}}\sqrt{1 + ({}_*\Delta_{a+\alpha-1}^{\alpha} y(t))^2} \to \min$$

$$y(a + \alpha - 1) = 0, \ y(b{+}\alpha - 1) = 1.$$

References

1. T. Abdeljawad, On Riemann and Caputo fractional differences, Comput. Math. Appl. **62** (2011), no. 3, 1602–1611.
2. T. Abdeljawad, On delta and nabla Caputo fractional differences and dual identities, Discrete Dyn. Nat. Soc. **2013**, Art. ID 406910, 12 pp.
3. S. Abe, A note on the q-deformation-theoretic aspect of the generalized entropies in nonextensive physics, Phys. Lett. A **224** (1997), no. 6, 326–330.
4. J. M. Amigó, S. G. Balogh and S. Hernández, A brief review of generalized entropies, Entropy **20** (2018), no. 11, Paper No. 813, 21 pp.
5. P. T. Anh et al., Variation of constant formulas for fractional difference equations, Arch. Control Sci. **28(64)** (2018), no. 4, 617–633.
6. F. M. Atici and P. W. Eloe, A transform method in discrete fractional calculus. Int. J. Difference Equ. 2 (2007), no. 2, 165–176.
7. F. M. Atici and P. W. Eloe, Initial value problems in discrete fractional calculus, Proc. Amer. Math. Soc. **137** (2009), no. 3, 981–989.
8. F. M. Atıcı and P. W. Eloe, Two-point boundary value problems for finite fractional difference equations, J. Difference Equ. Appl. **17** (2011), no. 4, 445–456.
9. F. M. Atıcı and S. Şengül, Modeling with fractional difference equations, J. Math. Anal. Appl. **369** (2010), no. 1, 1–9.
10. N. R. O. Bastos, R. A. C. Ferreira and D. F. M. Torres, Necessary optimality conditions for fractional difference problems of the calculus of variations, Discrete Contin. Dyn. Syst. **29** (2011), no. 2, 417–437.
11. J. Baoguo, The asymptotic behavior of Caputo delta fractional equations, Math. Methods Appl. Sci. **39** (2016), no. 18, 5355–5364.
12. M. Bohner, Calculus of variations on time scales, Dynam. Systems Appl. **13** (2004), no. 3-4, 339–349.
13. L. Boltzmann, Uber die Beziehung eines allgemeinen mechanischen Satzes zum zweiten Hauptsatz der Warmetheorie, Sitz. Ber. Akad. Wiss. Wien (II) 1877, **75**, 67–73.
14. E. P. Borges and I. Roditi; A family of non-extensive entropies, Phys. Lett. A **246**, 399 (1998).
15. X. Cao and S. Luo, On the stability of generalized entropies, J. Phys. A: Math. Theor. **42** (2009) 075205 (9pp).
16. J. Čermák, I. Győri and L. Nechvátal, On explicit stability conditions for a linear fractional difference system, Fract. Calc. Appl. Anal. **18** (2015), no. 3, 651–672.
17. R. Clausius, The Mechanical Theory of Heat; McMillan and Co.: London, UK, 1865.

R. A. C. Ferreira, *Discrete Fractional Calculus and Fractional Difference Equations*, SpringerBriefs in Mathematics,
https://doi.org/10.1007/978-3-030-92724-0

18. K. Diethelm, *The analysis of fractional differential equations*, Lecture Notes in Mathematics, 2004, Springer-Verlag, Berlin, 2010.
19. S. Elaydi, *An introduction to difference equations*, third edition, Undergraduate Texts in Mathematics, Springer, New York, 2005.
20. R. A. C. Ferreira, *Calculus of variations on time scales and discrete fractional calculus = Cálculo das Variações em Escalas Temporais e Cálculo Fraccionário Discreto*, ProQuest LLC, Ann Arbor, MI, 2010.
21. R. A. C. Ferreira, A discrete fractional Gronwall inequality, Proc. Amer. Math. Soc. **140** (2012), no. 5, 1605–1612.
22. R. A. C. Ferreira, Existence and uniqueness of solution to some discrete fractional boundary value problems of order less than one, J. Difference Equ. Appl. **19** (2013), no. 5, 712–718.
23. R. A. C. Ferreira, A new look at Bernoulli's inequality, Proc. Amer. Math. Soc. **146** (2018), no. 3, 1123–1129.
24. R. A. C. Ferreira, Fractional calculus of variations: a novel way to look at it, Fract. Calc. Appl. Anal. **22** (2019), no. 4, 1133–1144.
25. R. A. C. Ferreira and C. S. Goodrich, Positive solution for a discrete fractional periodic boundary value problem, Dyn. Contin. Discrete Impuls. Syst. Ser. A Math. Anal. **19** (2012), no. 5, 545–557.
26. R. A. C. Ferreira and J. T. Machado; An Entropy Formulation Based on the Generalized Liouville Fractional Derivative, Entropy **21**, 638 (2019).
27. R. A. C. Ferreira and D. F. M. Torres, Higher-order calculus of variations on time scales, in *Mathematical control theory and finance*, 149–159, Springer, Berlin.
28. R. A. C. Ferreira and D. F. M. Torres, Fractional h-difference equations arising from the calculus of variations, Appl. Anal. Discrete Math. **5** (2011), no. 1, 110–121.
29. I. M. Gelfand and S. V. Fomin, *Calculus of variations*, Revised English edition translated and edited by Richard A. Silverman, Prentice-Hall, Inc., Englewood Cliffs, NJ, 1963.
30. J.W. Gibbs, Elementary Principles in Statistical Mechanics–Developed with Especial References to the Rational Foundation of Thermodynamics; C. Scribner's Sons: New York, NY, USA, 1902.
31. C. S. Goodrich, Continuity of solutions to discrete fractional initial value problems, Comput. Math. Appl. **59** (2010), no. 11, 3489–3499.
32. C. S. Goodrich, On a first-order semipositone discrete fractional boundary value problem, Arch. Math. (Basel) **99** (2012), no. 6, 509–518.
33. C. S. Goodrich, On a fractional boundary value problem with fractional boundary conditions, Appl. Math. Lett. **25** (2012), no. 8, 1101–1105.
34. C. S. Goodrich, On a fractional boundary value problem with fractional boundary conditions, Appl. Math. Lett. **25** (2012), no. 8, 1101–1105.
35. C. Goodrich and A. C. Peterson, *Discrete fractional calculus*, Springer, Cham, 2015.
36. H. L. Gray and N. F. Zhang, On a new definition of the fractional difference, Math. Comp. **50** (1988), no. 182, 513–529.
37. T. H. Gronwall, Note on the derivatives with respect to a parameter of the solutions of a system of differential equations, Ann. of Math. (2) **20** (1919), no. 4, 292–296.
38. J. Henderson, Existence of local solutions for fractional difference equations with Dirichlet boundary conditions, J. Difference Equ. Appl. **25** (2019), no. 6, 751–756.
39. M. Holm, Sum and difference compositions in discrete fractional calculus, Cubo **13** (2011), no. 3, 153–184.
40. B. Jia, L. Erbe and A. Peterson, Comparison theorems and asymptotic behavior of solutions of discrete fractional equations, Electron. J. Qual. Theory Differ. Equ. **2015**, Paper No. 89, 18 pp.
41. W. G. Kelley and A. C. Peterson, *Difference equations*, second edition, Harcourt/Academic Press, San Diego, CA, 2001.
42. A. A. Kilbas, H. M. Srivastava and J. J. Trujillo, *Theory and applications of fractional differential equations*, North-Holland Mathematics Studies, 204, Elsevier Science B.V., Amsterdam, 2006.
43. B. Lesche, Instabilities of Rényi entropies, J. Statist. Phys. **27** (1982), no. 2, 419–422.

44. C. Lizama, The Poisson distribution, abstract fractional difference equations, and stability, Proc. Amer. Math. Soc. **145** (2017), no. 9, 3809–3827.
45. Sh. E. Mikeladze, De la résolution numérique des équations intégrales, Bull. Acad. Sci. URSS VII (1935), 255–257 (in Russian).
46. K. S. Miller and B. Ross, Fractional difference calculus, in *Univalent functions, fractional calculus, and their applications (Koriyama, 1988)*, 139–152, Ellis Horwood Ser. Math. Appl, Horwood, Chichester.
47. K. S. Miller and B. Ross, *An introduction to the fractional calculus and fractional differential equations*, A Wiley-Interscience Publication, John Wiley & Sons, Inc., New York, 1993.
48. L. J. Slater, *Generalized hypergeometric functions*, Cambridge University Press, Cambridge, 1966.
49. Tsallis, C; Possible generalization of Boltzmann–Gibbs statistics; J. Stat. Phys. **52**, 479 (1988).
50. J. G. Wendel, Note on the gamma function, Amer. Math. Monthly **55** (1948), 563–564.

Index

A
Asymptotically stable solution, 60

B
Bernoulli's inequality, 51
Binomial coefficient, 4

C
Chain rule, 5
Comparison result, 48
Convolution, 39

D
Discrete binomial theorem, 22
Discrete exponential function, 11
Discrete fractional calculus of variations, 71
du Bois-Reymond lemma, 72

E
Entropy, 35
Euler–Lagrange equation, 73

F
Falling function, 3
Forward difference operator, 1
Fractional chain rule, 32
Fractional difference power rule, 25
Fractional differences, 18
Fractional Gronwall's inequality, 49
Fractional Leibniz formula, 31
Fractional summation by parts, 21
Fractional summation power rule, 23
Fractional sums, 17
Fundamental lemma of the discrete calculus of variations, 72
Fundamental theorem of the discrete calculus, 7

G
Gronwall's inequality, 50

L
Legendre's necessary condition, 74
Leibniz formula, 4
Leibniz rule, 8
Liouville-Caputo fractional difference, 36

M
Mittag-Leffler function, 44

N
Natural boundary conditions, 79

R. A. C. Ferreira, *Discrete Fractional Calculus and Fractional Difference Equations*, SpringerBriefs in Mathematics,
https://doi.org/10.1007/978-3-030-92724-0

P
Pochhammer symbol, 4
Power rules, 3

S
Stable solution, 60
Summation by parts, 8
Summation operator, 6

V
Variation of constants formula, 11

W
Well-posedness, 63

Printed in the United States
by Baker & Taylor Publisher Services